Springer Tracts in Modern Physics
Volume 196

Available online at
SpringerLink.com

Starting with Volume 165, Springer Tracts in Modern Physics is part of the [SpringerLink] service. For all customers with standing orders for Springer Tracts in Modern Physics we offer the full text in electronic form via [SpringerLink] free of charge. Please contact your librarian who can receive a password for free access to the full articles by registration at:

www.springerlink.com

If you do not have a standing order you can nevertheless browse online through the table of contents of the volumes and the abstracts of each article and perform a full text search.

There you will also find more information about the series.

Springer
Berlin
Heidelberg
New York
Hong Kong
London
Milan
Paris
Tokyo

Physics and Astronomy

ONLINE LIBRARY

http://www.springer.de

Springer Tracts in Modern Physics

Springer Tracts in Modern Physics provides comprehensive and critical reviews of topics of current interest in physics. The following fields are emphasized: elementary particle physics, solid-state physics, complex systems, and fundamental astrophysics.

Suitable reviews of other fields can also be accepted. The editors encourage prospective authors to correspond with them in advance of submitting an article. For reviews of topics belonging to the above mentioned fields, they should address the responsible editor, otherwise the managing editor. See also www.springer.de

Managing Editor

Gerhard Höhler

Institut für Theoretische Teilchenphysik
Universität Karlsruhe
Postfach 69 80
76128 Karlsruhe, Germany
Phone: +49 (7 21) 6 08 33 75
Fax: +49 (7 21) 37 07 26
Email: gerhard.hoehler@physik.uni-karlsruhe.de
www-ttp.physik.uni-karlsruhe.de/

Elementary Particle Physics, Editors

Johann H. Kühn

Institut für Theoretische Teilchenphysik
Universität Karlsruhe
Postfach 69 80
76128 Karlsruhe, Germany
Phone: +49 (7 21) 6 08 33 72
Fax: +49 (7 21) 37 07 26
Email: johann.kuehn@physik.uni-karlsruhe.de
www-ttp.physik.uni-karlsruhe.de/~jk

Thomas Müller

Institut für Experimentelle Kernphysik
Fakultät für Physik
Universität Karlsruhe
Postfach 69 80
76128 Karlsruhe, Germany
Phone: +49 (7 21) 6 08 35 24
Fax: +49 (7 21) 6 07 26 21
Email: thomas.muller@physik.uni-karlsruhe.de
www-ekp.physik.uni-karlsruhe.de

Fundamental Astrophysics, Editor

Joachim Trümper

Max-Planck-Institut für Extraterrestrische Physik
Postfach 16 03
85740 Garching, Germany
Phone: +49 (89) 32 99 35 59
Fax: +49 (89) 32 99 35 69
Email: jtrumper@mpe-garching.mpg.de
www.mpe-garching.mpg.de/index.html

Solid-State Physics, Editors

Andrei Ruckenstein
Editor for The Americas

Department of Physics and Astronomy
Rutgers, The State University of New Jersey
136 Frelinghuysen Road
Piscataway, NJ 08854-8019, USA
Phone: +1 (732) 445 43 29
Fax: +1 (732) 445-43 43
Email: andreir@physics.rutgers.edu
www.physics.rutgers.edu/people/pips/
Ruckenstein.html

Peter Wölfle

Institut für Theorie der Kondensierten Materie
Universität Karlsruhe
Postfach 69 80
76128 Karlsruhe, Germany
Phone: +49 (7 21) 6 08 35 90
Fax: +49 (7 21) 69 81 50
Email: woelfle@tkm.physik.uni-karlsruhe.de
www-tkm.physik.uni-karlsruhe.de

Complex Systems, Editor

Frank Steiner

Abteilung Theoretische Physik
Universität Ulm
Albert-Einstein-Allee 11
89069 Ulm, Germany
Phone: +49 (7 31) 5 02 29 10
Fax: +49 (7 31) 5 02 29 24
Email: frank.steiner@physik.uni-ulm.de
www.physik.uni-ulm.de/theo/qc/group.html

Christian Binek

Ising-type Antiferromagnets

Model Systems in Statistical Physics
and in the Magnetism of Exchange Bias

With 52 Figures

 Springer

Priv.-Doz. Dr. Christian Binek
Gerhard-Mercator-Universität Duisburg
Fakultät für Naturwissenschaften
Institut für Physik
Laboratorium für Angewandte Physik
Lotharstraße 1
47057 Duisburg, Germany
E-mail: binek@kleemann.uni-duisburg.de

Cataloging-in-Publication Data applied for

A catalog record for this book is available from the Library of Congress.

Bibliographic information published by Die Deutsche Bibliothek

Die Deutsche Bibliothek lists this publication in the Deutsche Nationalbibliografie; detailed bibliographic data is available in the Internet at http://dnb.ddb.de.

Physics and Astronomy Classification Scheme (PACS):
64.10.+h, 75.50.Ee, 75.70.Cn, 75.70.Ak

ISSN print edition: 0081-3869
ISSN electronic edition: 1615-0430
ISBN 3-540-40428-7 Springer-Verlag Berlin Heidelberg New York

Springer-Verlag Berlin Heidelberg New York
a member of BertelsmannSpringer Science+Business Media GmbH

http://www.springer.de

© Springer-Verlag Berlin Heidelberg 2003
Printed in Germany

Typesetting: Camera-ready copy from the author using a Springer LATEX macro package
Production: LE-TEX Jelonek, Schmidt & Vöckler GbR, Leipzig
Cover concept: eStudio Calamar Steinen
Cover production: *design &production* GmbH, Heidelberg

Printed on acid-free paper   SPIN: 10910191   56/3141/YL   5 4 3 2 1 0

Preface

The early history of magnetism started with the discovery of the natural mineral magnetite named after the Asian province Magnesia. Since that time, the magnetism of solids preserved its lasting fascination. This holds for the phenomenon itself as well as for its various applications. Among them the Chinese compass represents the most famous historical magnetic device, which has been known for more than 3000 years.

Although solid-state magnetism has attracted people for such a long time, its understanding is still a challenging task which requires elements of the most fascinating fields of modern physics. For instance, the Bohr van-Leeuwen theorem rigorously evidences the pure quantum nature of magnetism. The relativistic effect of spin orbit coupling is a substantial ingredient in order to understand magnetic anisotropy, which is of major importance for most applications. Moreover, the understanding of the thermodynamic properties of magnetic materials requires sophisticated methods of modern statistical physics. It is therefore not surprising that the investigation of cooperative and critical magnetic phenomena gives rise to significant progress in statistical physics.

A similar situation holds for the mutual relationship between magnetism and the development of experimental methods. For example, modern state of the art magnetometry benefits from superconducting quantum interference devices, which are based on Josephson's macroscopic quantum effects. Huge progress in spin-dependent scattering, diffraction and absorption techniques of neutrons, electrons and X-rays stimulate the investigation of magnetic structures and excitations in bulk and artificially structured materials. Nonlinear optical methods became sensitive to the broken time and spatial inversion symmetry of magnetic domain states and buried interfaces, respectively. In addition, scanning probe microscopy entered the field of micromagnetism via magnetic force microscopy, while spin-polarized scanning tunnelling techniques are able to image magnetic surfaces down to the atomic scale.

However, this breathtaking progress cannot obscure the fact that, up to now, research activities have focused mainly on ferromagnetic materials while more complex types of magnetic order have attracted less interest. This imbalance has two obvious reasons. On the one hand, the ferromagnetic order-parameter is easily accessible by magnetometry and, on the other hand, most applications require the presence of a macroscopic permanent magnetic moment. Historically, the development of neutron-scattering techniques initiated at least a growing scientific interest in more complex magnetic structures when they changed from an abstract theoretical hypothesis into experimental facts. In particular, antiferromagnetism with its variety of complex spin arrangements became more and more popular. Nowadays, collinear antiferromagnets are among the most prominent magnetic model systems and play a crucial role in the physics of critical phenomena. In particular, insulating ionic compounds with localized magnetic moments represent perfect realizations of anisotropic Heisenberg systems. As a rule, antiferromagnetic superexchange between the cations is mediated via the orbitals of the anions by charge-transfer processes. Crystal field components with non-cubic symmetry determine the single-ion anisotropy, which in particular cases favors uniaxial spin orientation of the Ising type. Ising systems, however, play an outstanding role in statistical physics and their experimental realizations are of major importance.

The first part of this monograph deals with the prototypical ionic compound $FeCl_2$ which is one of the rare realizations of 2d Ising systems. Interestingly and typically, its low-dimensional ferromagnetic behavior is replaced by 3d antiferromagnetic order on cooling to below the critical temperature. Here, the weak antiferromagnetic interaction drives the system into the long range ordered state and prevents criticality of the ferromagnetic fluctuations. However, above the Néel temperature precise measurements of the isothermal magnetization allow for experimental access to the statistical theory of Lee and Yang. Starting with a brief introduction of this theory from an experimentalists point of view, the analysis of the data is presented and density functions of the complex zeros of the partition function as well as the gap exponent for the 2d Ising ferromagnet on a triangular lattice are determined and compared with theoretical results.

Very recently, antiferromagnetism pushed further into the center of research interests when it turned out that, in many cases, the technical progress of magnetic devices originates from the impact of optimized antiferromagnetic materials. For example, novel concepts of magnetic data-storage devices involve antiferromagnetic components in order to overcome the superparamagnetic limit. This technological step becomes necessary when scaling down of the ferromagnetic representation of a bit gives rise to thermally activated magnetization reversal on an unacceptable time scale. Moreover, there are new artificial heterostructures of antiferromagnetic and ferromagnetic thin films which recently came into focus of the field of spinelectronics. Although problems like, e.g., the efficient injection of polarized spins into a semiconductor still prevent the realization of a spintransistor, the application of the exchange bias effect in spinvalves, fieldsensors and non-volatile magnetic random-access memories is a promising starting point for the emerging spinelectronic. Here antiferromagnets play an important role as pinning layers for the adjacent ferromagnetic thin films of the exchange-bias heterostructures, which are basic components of the corresponding passive spin-electronic devices.

The second part of this book deals with the exchange-bias phenomenon in prototypical magnetic heterostructures. Special emphasis is laid on model systems where the limited number of spin degrees of freedom minimizes the complexity of the problem. A generalization of the phenomenological Meiklejohn Bean approach is presented which takes into account, e.g., finite anisotropy and thickness of the antiferromagnetic pinning layer. Mechanisms which create the antiferromagnetic interface magnetization are investigated and discussed. Special attention is paid to piezomagnetism. This is a well-known bulk phenomenon, e.g., in the case of rutile-type antiferromagnets, but has been overlooked so far as a significant contribution to the antiferromagnetic interface magnetization of exchange-bias heterostructures.

This book is based essentially on work submitted in partial fulfillment of the requirements of my Habilitation at the former Gerhard-Mercator-Universität Duisburg, now being adopted into the University Duisburg-Essen. It is a great pleasure for me to express my sincere gratitude to W. Kleemann who offered me the essential possibilities for the development of my scientific career. It is obvious that this fruitful collaboration in recent years, e.g., within the framework of the project A7 'Spin structures and exchange bias of magnetic metal insulator heterosystems' of the Sonderforschungsbereich 491 'Magnetic Heterostructures: Structure and Electronic Transport' additionally required the work of numerous colleagues. I would like to deeply thank all of them for their stimulating work in a friendly atmosphere. I would also like to thank the German Science Foundation (DFG) for its kind support through SFB 491 and the Graduate School 277 'Structure and Dynamics of Heterogeneous Systems'.

Duisburg,
August 2003   *Christian Binek*

Contents

1 Introduction

It is the aim of this book to shed light on selected modern aspects of artificially layered structures and bulk materials involving antiferromagnetic long-range order. Special emphasis is laid on the prototypical behavior of Ising-type model systems. They play an outstanding role in the field of statistical physics and provide experimental access to the theory of Lee and Yang. In addition, basic mechanisms of the exchange-bias effect are studied with the help of magnetic heterosystems involving Ising-type antiferromagnets. Their minimized number of spin degrees of freedom gives rise to model-type behavior. This approach allows to reduce the complexity of the phenomenon and provides access to the basic mechanisms of exchange bias.

The current interest in magnetism and the abundance of activities in this field is based on a few unifying aspects. Traditionally, the phenomenon of cooperative magnetic order in bulk materials is one of the most important branches of statistical physics in general and critical phenomena in particular [1, 2, 3]. The reason for this outstanding role of magnetism is given by the fact that predictions about universal properties of complex systems require information about the spatial dimensionality as well as the symmetry of the interaction. Magnetic model systems provide an easy access to these parameters and are, hence, the trailblazers in the physics of critical phenomena.

Antiferromagnets play a particular role in the experimental study of critical behavior. First of all, the temperature-driven transition from the paramagnetic into the antiferromagnetic (AF) phase maintains its criticality in the presence of moderate magnetic fields. This is an advantage with respect to ferromagnetic (FM) transitions where, on the one hand, magnetic fields are often needed in order to avoid domain effects but, on the other hand, the conjugate field destroys the critical behavior. Secondly, model systems with localized magnetic moments are best realized by insulating ionic compounds, where as a rule AF superexchange interaction is involved [4]. Very frequently, owing to an internal hierarchy of interactions, such systems offer AF and FM behavior when the thermal energy approaches the energy of the respective magnetic interaction. It is therefore not surprising that ionic compounds with direct FM exchange and additional weak AF interaction are among the best available model systems of low-dimensional (d) Ising ferromagnets. For instance, far above the Néel temperature of $FeCl_2$, the thermodynamic behavior of this 3d antiferromagnet is dominated by a strong FM in-plane interaction between the effective Ising spins of the Fe^{2+} ions. Hence, this simple compound possesses various prototypical properties. On the one hand, it is a model system of metamagnetism. On the other hand, the dominating in-plane interaction makes $FeCl_2$ one of the rare realizations of a 2d Ising ferromagnet on a triangular lattice.

Experiments on this system supply information about basic thermodynamic properties. For example, the analysis of its isothermal magnetization provides experimental access to the statistical theory of Lee and Yang [5]. In their profound analysis of the Ising model, they discovered a fundamental rigorous result which is known as the 'circle theorem'. It predicts that the zeros of the partition function of an Ising ferromagnet or a lattice gas are distributed on the unit circle in the complex magnetic field plane. The analysis of experimental magnetization data provides access to the distribution function of the zeros on the unit circle. Even the exotic critical behavior, which Ising ferromagnets exhibit in purely imaginary magnetic fields, can be analyzed from the power-law divergence of the zero-density functions.

In addition to magnetic bulk properties, a new and fascinating branch in magnetism is stimulated by modern techniques of nanometric structuring and multilayer growth. These novel techniques enable the fabrication of new artificial materials. For example, insulating AF and itinerant FM materials are combined in order to design heterostructures with novel magnetic properties. Although such compositions can be far from thermodynamic equilibrium, they are stable on a long time scale and exhibit new, interface-induced magnetic features like, for example, the exchange-bias effect. In addition to static proximity effects, such multilayer systems reveal exciting new transport properties. For instance, giant magnetoresistance (GMR) and tunneling magnetoresistance (TMR) effects arise in exchange-coupled ferromagnets which are separated by thin non-magnetic and isolating spacer layers, respectively. Both effects are of considerable technological relevance, but still far from being completely understood.

Besides these branches of basic research interest, modern magnetism provides a huge potential of applications. In many cases, the technical progress originates from the physical impact of antiferromagnetism. It manifests, for example, in the rapid development of magnetic data-storage devices. The latest generation of modern hard disks is fabricated from artificial antiferromagnets consisting of AF-coupled FM thin films in order to overcome the paramagnetic limit of single-domain particles [6, 7]. Moreover, there is a growing interest in AF materials which originates from the application of the exchange-bias effect in spin valves, field sensors and non-volatile magnetic random access memories (M-RAMs). Such devices, which combine a static interface with electronic transport properties, are at the basis of the very promising future technology of spin electronics [8, 9, 10].

Exchange bias plays an important role in modern passive spin-electronic devices. It represents a coupling phenomenon between FM and AF materials which is phenomenologically characterized by a shift of the FM hysteresis loop along the magnetic field axis. This shift reflects a unidirectional anisotropy which originates from the interface coupling of the ferromagnet and its AF pinning layer. Although the exchange-bias effect has been known for 45 years, its physical basis is still under debate.

In order to understand this phenomenon on a microscopic level, detailed insight into the complex thermal and field-dependent evolution of the spin structure of the heterosystem is needed. In order to tackle this complex problem, model systems are required, which exhibit the basic mechanisms of the exchange-bias effect. Heterolayer structures of uniaxially anisotropic antiferro- and ferromagnets involve a limited number of spin degrees of freedom. Their minimized complexity of possible spin arrangements facilitates the model-type behavior of such heterosystems.

These various aspects of uniaxial antiferromagnets make them a unique subject in magnetism. It is the aim of this work to point out some of these fundamental as well as applied aspects of the physics of antiferromagnetism.

References

1. C. Domb and M.S. Green: *Phase Transitions and Critical Phenomena, Vol. 1* (Academic, London 1972)
2. H.E. Stanley: *Introduction to Phase Transitions and Critical Phenomena* (Oxford University Press, New York 1971)
3. W. Gebhard and U. Krey: *Phasenübergänge und kritische Phänomene* (Vieweg, Braunschweig 1980)
4. L.J. De Jongh and A.R. Miedema: Adv. Phys. **23**, 1 (1974)
5. T.D. Lee and C.N. Yang: Phys. Rev. **87**, 410 (1952)
6. E. Fullerton, D.T. Margulies, M.E. Schabes, M. Carey, B. Gurney, A.Moser, M. Best, G. Zeltzer, K. Rubin, H. Rosen and M. Doerner: Appl. Phys. Lett. **77**, 3806 (2000)
7. S. Jorda, H. Kock and M. Rauner: Phys. Bl. **57**, 18 (2001)
8. St. Mengel: XMR-Technologien. In: *Physikalische Technologien, Band 20* (VDI-Technologiezentrum Physikalische Technologien,Düsseldorf 1977)
9. G. Prinz and K. Hathaway: Phys. Today **4**, 24 (1995)
10. S. Datta and B. Das: Appl. Phys. Lett. **56**, 665 (1990)

Part I

Model Systems in Statistical Physics

2 Ising-type Antiferromagnets: Model Systems in Statistical Physics

This chapter points out the impact of insulating ionic compounds with localized magnetic moments in the field of statistical physics. They are perfect realizations of anisotropic Heisenberg systems which, in the case of strong single-ion anisotropy, enable the realization of Ising-type model systems. Such compounds involving strong direct ferromagnetic exchange and additional weak antiferromagnetic interaction turn out to be among the best available model systems of low-dimensional Ising ferromagnets. $FeCl_2$ is a representative of these prototypical systems which provides experimental access to fundamental thermodynamic properties in general and the statistical theory of Lee and Yang in particular. Starting with a brief introduction of the Lee–Yang theory from an experimentalist's point of view, density functions of the complex zeros of the partition function as well as the gap exponent of the 2d Ising ferromagnet on a triangular lattice are determined from isothermal magnetization data. Finally, modern approaches beyond the Lee–Yang theory are briefly discussed. Fisher's zeros in the complex temperature plane are introduced. A possible simple basis underlying the circle theorem is motivated and implications from the Lee–Yang theory in the field of non-equilibrium systems are briefly discussed.

2.1 Uniaxially Anisotropic Antiferromagnets with Localized Moments

Magnetic systems of interacting localized moments are of particular interest in the framework of statistical physics. This originates from the fact that their magnetic properties can be mapped onto classical spin systems which enable a straightforward computation of the total energy of a given spin configuration. Hence, thermal averages can be calculated in some cases by analytical and, more generally, by numerical methods like Monte Carlo simulations. In the case of itinerant magnetism, however, the determination of the ground-state energy is already a difficult task. It requires ab initio calculations which are based on the density functional approach or other involved methods of many particle-physics [1]. Moreover, in order to get access to thermodynamic quantities, knowledge about the energy spectra of the low-lying excitations is required. Once the relevant energies are found the problem can be mapped onto a classical spin model and common methods of statistical physics become applicable [2]. This outstanding role of classical spin models motivates the experimental investigation of magnetic systems which obey the anisotropic Heisenberg model. A fruitful interplay between experimental and theoretical investigations is necessary in order to work out the minimum set of parameters which are required for a consistent interpretation of the experimental findings in model systems like, e.g., $FeBr_2$ [3, 4, 5, 6, 7, 8, 9, 10].

The magnetic properties of insulating antiferromagnets of the MX_2 type (M = Mn, Fe, Co, Ni and X = F, Cl, Br) are perfect realizations of various anisotropic Heisenberg models [11, 12]. The ionic character of the chemical bonding ensures the localization of the magnetic moments in a nearly ideal manner. The magnetic properties of the FeX_2 compounds with X = Cl and Br, respectively, originate from a common structure of the underlying Hamiltonian. It is based on the nearly octahedral symmetry of the crystal field which acts on the Fe^{2+} ions and determines their energy level structure [13]. Hence, the corresponding electrostatic potential is invariant under symmetry operations of the O_h point group, a subgroup of the 3d full rotation group of the free ion [14]. In accordance with Hund's rules, the electronic $3d^6$ state of the free Fe^{2+} ion yields the total orbital and spin angular momentum $L = 2$ and $S = 2$, respectively. The crystal field-induced lowering of the symmetry lifts the five-fold L-degeneracy of the ^5D state of the free ion. It splits into a triplet $^5\Gamma_{2g}$ and the doublet ^5E_g state. The energetically low lying triplet is subject to additional perturbations which arise from spin-orbit coupling , a residual trigonal component of the crystal field and the exchange interaction [13]. These perturbations are very small in comparison with the strong cubic crystal field. Hence,$^5\Gamma_{2g}$ is described by an effective ^5P state where the orbital angular momentum $L = 1$ has partly been quenched. The perturbation which arises from the spin orbit coupling lifts the degeneracy of the triplet state. The resulting total angular momentum $\boldsymbol{J} = \boldsymbol{L} + \boldsymbol{S}$ provides three states which are labeled by the quantum numbers $J = 1, 2$ and 3 [15]. The $J = 1$ ground state determines the main magnetic properties of the FeX_2 compounds. The energetic separation of the $J = 1$ multiplet from the $J = 2$ and $J = 3$ states justifies the introduction of an effective spin with quantum number $S = 1$ [15]. Within the $S = 1$ multiplet the effective spin Hamiltonian reads

$$\hat{H} = -\sum_{i,j} \left(J_{ij}\, \hat{\boldsymbol{S}}_i \hat{\boldsymbol{S}}_j + K_{ij}\, \hat{S}_i^z\, \hat{S}_j^z \right) - D \sum_i \left(\hat{S}_i^z \right)^2 - g\,\mu_B\mu_0 H \sum_i \hat{S}_i^z.$$

(2.1)

The i, j summation takes into account all Fe^{2+} sites of the crystal. The exchange interaction between two spins $\boldsymbol{S}_i$ and $\boldsymbol{S}_j$ contains an isotropic interaction of the strength J_{ij} and an anisotropic term involving the exchange constant K_{ij}. While negative exchange constants favor AF spin ordering, positive exchange favors parallel alignment of the spins. The existence of anisotropic exchange is a consequence of the effective-spin approach. It neglects the small remaining interference between $J = 1$ and higher J states.

In order to compensate this approximation, the former isotropic exchange between the total angular momentum J of the Fe^{2+} ions is projected onto the effective anisotropic exchange interaction between effective spins.

The second sum in (2.1) originates from the additional trigonal component of the crystal field. It creates the single-ion anisotropy with its corresponding energy contribution [15]. The D parameter tunes the strength of the anisotropy and, hence, controls the Ising-type character of the system. Large positive D values favor parallel or antiparallel orientation of the spin along the z direction. The last term of the Hamiltonian represents the Zeeman energy of the magnetic moments in the presence of a magnetic field $\mu_0 H$ which points along the z direction. The effective g factor and the Bohr magneton, μ_B, are proportionality constants between the spin and its magnetic moment.

Within the next paragraph, the role of $FeCl_2$ as an Ising-type model system is stressed. Hence, the microscopic exchange and anisotropy parameters of this compound are introduced in detail. Figure 2.1 exhibits the crystalline structure of $FeCl_2$. x_r, y_r, z_r indicate the basis vectors of the rhombohedral unit cell. It builds up a lattice of space group D_{3d}^5 symmetry [16]. Hexagonal layers of Fe^{2+} ions (solid circles) are separated by two layers of Cl^- ions (open circles). Within the hexagonal layers isotropic FM interaction of the strength $J_1/k_B = 3.9$ K takes place between six nearest-neighbors. Small negative isotropic exchange, $J_2/k_B = -0.52$ K, gives rise to weak AF coupling between next-nearest-neighbors within the Fe^{2+} layers. Additional anisotropic intralayer exchange , $K/k_B = -2.2$ K, is limited to nearest-neighbors. The separation of the Fe^{2+} layers by two Cl^- layers gives rise to weak AF superexchange interlayer coupling, $J/k_B = -0.18$ K. In accordance with the large distance $c/3 = 0.585$ nm between adjacent Fe^{2+} layers, the magnitude of J is quite small in comparison with the FM intralayer coupling J_1. Nevertheless, the 3d AF long-range order occurring at $T < T_N = 23.7$ K originates from this small AF interlayer exchange.

Table 2.1 summarizes the microscopic parameters which enter the Hamiltonian. The values of the in-plane interaction constants are based on the analysis of the planar spin-wave spectra [13]. The AF inter layer exchange can be determined from the metamagnetic spin-flip field according to $g\mu_B H = 2z|J'|$, where z is the number of nearest-neighbors [11]. Usually, and in accordance with Table 2.1, z is determined by counting the number of geometrical nearest-neighbors between adjacent Fe^{2+} layers. However, on taking into account the equivalence of distinct paths of superexchange, the number of nearest-neighbors increases from six to 24 for both adjacent layers. Hence, the corresponding interlayer exchange constant decreases by a factor of four.

2.2 Lee–Yang Theorem

In 1952 Lee and Yang [17] pointed out that the distribution of the zeros of the partition function Z of an Ising ferromagnet reveals a remarkable symmetry

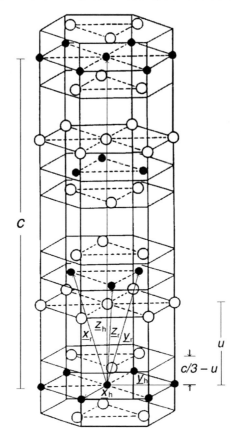

Fig. 2.1. Crystalline structure of FeCl$_2$. x_h, y_h, z_h and x_r, y_r, z_r denote the hexagonal and rhombohedral basis vectors of the corresponding unit cells. *Solid* and *open circles* represent Fe^{2+} and Cl$^-$ ions, respectively. $c = 1.7536$ nm is the length of the c axis and $a = 0.3579$ nm is the length of the basis vectors within a hexagonal layer. The distance between adjacent Fe^{2+} and Cl$^-$ layers reads $c/3 - u$, where $u = 0.2543c$.

Table 2.1. Parameters of the Hamiltonian (2.1) describing the magnetic properties of FeCl$_2$

$S = 1$	effective spin quantum number
$J_1/k_B = 3.9$ K	isotropic in-plane exchange of nearest-neighbors
$J_2/k_B = -0.52$ K	isotropic in-plane exchange of next-nearest-neighbors
$K/k_B = -2.2$ K	anisotropic in-plane exchange of nearest-neighbors
$J/k_B = -0.18$ K	isotropic inter layer exchange of nearest-neighbors
$D/k_B = 9.8$ K	single-ion anisotropy
$g = 4.1$	effective g value

in the complex fugacity plane. By virtue of their famous theorem they proved that, in the case of an Ising ferromagnet, the zeros of Z are distributed on the unit circle $z = \exp(i\theta)$ in the complex $z = \exp(-2gS\mu_B\mu_0H/k_BT)$ plane, where $gS\mu_B$ is the magnetic moment with Bohr's magneton μ_B, the spin quantum number S and the Landé factor g, while H is the complex magnetic field and T is the temperature. The starting point of their theory is the Ising Hamiltonian

$$\tilde{H} = -\sum_{i,j} J_{ij}s_is_j - h\sum_i s_i, \qquad (2.2)$$

where $h = g\mu_B\mu_0H$ and $S_i = \pm 1$ is the classical Ising-spin variable. The Ising Hamiltonian can be derived from the classical anisotropic Heisenberg expression, (2.1), in the limit of infinite positive single-ion anisotropy. This fact becomes of major importance when looking for experimental realizations of (2.2). The corresponding partition function for N Ising spins which interact according to (2.2) reads

$$Z = e^{N\beta h}\sum_{\{s\}} e^{\beta(\sum_{i,j} J_{ij}s_is_j + h(\sum_i s_i - N))}, \qquad (2.3)$$

where $\beta = 1/k_BT$ and $\{s\}$ denotes the summation with respect to all possible spin configurations. Note that $\sum_i S_i - N$ is an even number for every configuration of $\{s\}$ and is bounded from below and above according to $-2N \leq \sum_i S_i - N \leq 0$. Hence, Z is a polynomial of order N in the variable $z = \exp(-2\beta h)$. It reads

$$Z = e^{N\beta h}\sum_{n=0}^{N} K_n z^n, \qquad (2.4)$$

with positive coefficients K_n which do not depend on the magnetic field.

An arbitrary polynomial of the order N has N complex zeros which are in general irregularly distributed in the complex plane. However, the zeros of the partition function of the Hamiltonian (2.2) are located on the unit circle in the case of FM coupling constants. This remarkable symmetry is the essence of the Lee–Yang theorem. Note that, in the case of finite systems, (2.4) represents a sum of a finite number of positive terms. Hence, there is no real magnetic field that sets Z to zero. Therefore, all thermodynamic quantities remain analytic. In the thermodynamic limit of infinite system size, however, the complex zeros can touch the positive real axis of the complex z plane and phase transitions become possible [18].

The important implications of the celebrated Lee–Yang circle theorem can be illustrated by the following elementary example of two interacting Ising spins in a field. Their Hamiltonian reads

$$\tilde{H} = -Js_1s_2 - h(s_1 + s_2). \qquad (2.5)$$

The four spin configurations involving two antiparallel, one parallel up and one parallel down alignment of the spins yields four energy values with a two-fold degeneracy of the two states with zero total spin. The corresponding Boltzmann factors build up the partition function

$$Z = e^{\beta(J+2h)} + 2e^{-\beta J} + e^{\beta(J-2h)}. \tag{2.6}$$

Substitution of $exp(-2\beta h) = z$ into (2.6) yields the polynomial

$$Z = e^{2\beta h}e^{\beta J}\left(z^2 + 2e^{-2\beta J}z + 1\right). \tag{2.7}$$

Its rearrangement into the product representation reads

$$Z = e^{2\beta h}e^{\beta J}\left(z^2 + 2e^{-2\beta J}z + 1\right), \tag{2.8}$$

where the complex zeros $\xi_{1/2}$ are given by

$$\xi_{1/2} = -e^{-2\beta J} \pm i\sqrt{1 - e^{-4\beta J}}. \tag{2.9}$$

As long as the prerequisite of FM interaction $J > 0$, is fulfilled, the complex zeros obey the condition $\left|\xi_{1/2}\right| = 1$ and, hence, lie on the unit circle. However, in the case $J < 0$, the $\xi_{1/2}$ lie on the negative real axis and are separated from the unit circle. Figure 2.2 illustrates this behavior in the case of $\beta J = 0.5, 0.1, 0$ and -0.1, respectively.

2.3 Lee–Yang Zeros and Thermodynamics

The circle theorem points out that all complex zeros of the partition function of Ising ferromagnets are located on the unit circle in the complex z plane. Remarkably, this symmetry is independent of the lattice dimension. However, no information about the precise distribution and thermal evolution of the zeros is provided from the theorem. This cannot be expected, because access to this information is equivalent to the complete knowledge of the thermodynamic behavior of the system. This problem is not yet solved in dimensions higher than two.

The intimate connection between the zeros and the thermodynamic behavior of a system becomes obvious after substitution of the product representation of the partition function into $F = -k_{\rm B}T\ln Z$, the basic relation between the free-energy and the partition function of the canonical ensemble. For N spins the free-energy reads

$$F = -Nh - k_{\rm B}T\sum_{i=1}^{N}\ln(z - \xi_i). \tag{2.10}$$

In the thermodynamic limit of an infinite particle number, the discrete distribution of the $\xi_k = e^{i\theta_k}$ becomes a continuous zero density function $g(\theta)$,

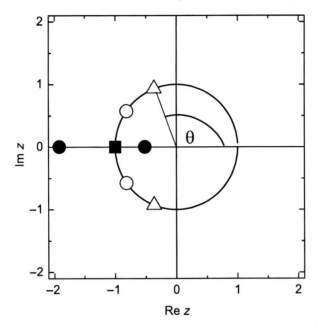

Fig. 2.2. Positions of the zeros $\xi_{1/2}$ in the complex z plane for $\beta J = 0.5$ (*triangles*), 0.1 (*open circles*), 0 (*square*) and -0.1 (*solid circles*), respectively. θ denotes the polar angle of the Euler representation of ξ. In the case $\beta J \geq 0$ the zeros lie on the unit circle, while $\beta J = -0.1$ yields zeros inside and outside the circle on the negative real axis

where $g(\theta)$ is the number of zeros between θ and $\theta + \mathrm{d}\theta$. Replacing the summation in (2.10) by integration with respect to θ yields

$$\frac{F}{N} = -h - k_{\mathrm{B}}T \int_0^{2\pi} g(\theta) \ln(z - \mathrm{e}^{\mathrm{i}\theta})\mathrm{d}\theta, \qquad (2.11)$$

where the zero-density function fulfills the constraint

$$\int_0^{2\pi} g(\theta)\mathrm{d}\theta = 1. \qquad (2.12)$$

The free-energy and its derivatives are real quantities. Hence, every complex zero $\xi_k = \mathrm{e}^{\mathrm{i}\theta_k}$ must be accompanied by a complex-conjugate partner $\xi_k = \mathrm{e}^{-\mathrm{i}\theta_k}$. Therefore, $g(\theta)$ is an even function of θ and (2.12) is transformed into

$$\frac{F}{N} = -h - k_{\mathrm{B}}T \int_0^{\pi} g(\theta) \ln(z^2 - 2z\cos\theta + 1)\mathrm{d}\theta, \qquad (2.13)$$

an explicitly real expression. Equation (2.13) allows a straightforward calculation of thermodynamic quantities. For example, the normalized magnetization is derived from (2.13) according to $I(z) = -1/N \ \partial F/\partial h$. One obtains

$$I(z) = 1 - 4z \int_0^\pi g(\theta) \frac{z - \cos\theta}{z^2 - 2z\cos\theta + 1} d\theta. \tag{2.14}$$

From a theoretical point of view, there are two approaches to investigate $g(\theta)$. On the one hand, straightforward solutions of $Z(z) = 0$ for dimensions $d \geq 2$ are possible, but restricted to systems of a few interacting spins only [19, 20, 21, 22]. On the other hand, the basic relation (2.14), which correlates I and $g(\theta)$, is used in order to approximate $g(\theta)$ from corresponding approximations of $I(z)$. The latter approach can be modified in order to obtain experimental access to $g(\theta)$, because $I(z)$ can be measured with very high precision, e.g. by superconducting quantum interference device (SQUID) techniques. Equation (2.14) is, hence, a preferred candidate for the experimental determination of the density function. However, in order to extract $g(\theta)$ from magnetization data, an explicit expression of the density function is required. It can be derived on substituting the Z transform of $\cos n\theta$ [23]

$$\sum_{n=0}^\infty \cos n\theta \ \tilde{z}^{-n} = \frac{\tilde{z}(\tilde{z} - \cos\theta)}{(\tilde{z}^2 - 2\tilde{z}\cos\theta + 1)} \tag{2.15}$$

into (2.14). Using the point symmetry of the magnetization with respect to the magnetic field, $I(1/z) = -I(z)$, one obtains

$$I(z) = 1 + 2\pi \sum_{n=1}^\infty g_n z^n \text{ with } g_n = \frac{2}{\pi} \int_0^\pi g(\theta) \cos n\theta d\theta. \tag{2.16}$$

Here g_n is the nth Fourier coefficient of the cosine series of $g(\theta)$. Equation (2.16) can be interpreted as the high-field series expansion of the magnetization with expansion coefficients g_n. In principle, a multi-parameter fit provides access to the leading g_n coefficients which build up the Fourier series of $g(\theta)$. Unfortunately, the poor convergence of (2.16) requires a huge number of expansion coefficients in order to satisfactorily represent $g(\theta)$. However, Kortman and Griffiths [24] pointed out that $g(\theta)$ may also be constructed from quickly converging Padé approximants of $I(z)$ when using the relation

$$g(\theta) = \frac{1}{2\pi} \lim_{r \to 1^-} \text{Re} \ I(r \exp(i\theta)). \tag{2.17}$$

The latter is easily verified by substitution of $z = r \exp(i\theta)$ into the high-field series from above. Equation (2.17) represents the basic relation for the analysis of the experimental magnetization data. Starting from high-field series expansions of $I(z)$ [25], Kortman and Griffiths computed Padé approximants and calculated $g(\theta)$ from (2.17) for Ising ferromagnets on a 2d square and a 3d diamond lattice at $T \neq T_c$. In addition, they investigated $g(\theta)$ for the mean-field model and the linear chain. Their solution of the latter problem was in good agreement with the rigorous analytical expression obtained previously by Lee and Yang [17].

In analogy to the procedure introduced by Kortman and Griffiths [24], $g(\theta)$ can be determined from m vs. H data by using (2.17) [26]. To this end, the data sets are normalized with respect to the low-temperature and high-field saturation value m_s and subsequently best-fitted to empirical functions of the type

$$f(z) = \frac{1 + n_1 z - (1 + n_1)z^2}{1 + d_1 z + d_2 z^2} \qquad (2.18)$$

with appropriate parameters n_1, d_1 and d_2. The fitting functions take into account the limiting cases $f(z = 1) = 0$ and $f(z = 0) = 1$ of the normalized magnetization at $T > T_c$ in zero and infinite magnetic field, respectively. Once the fitting parameters of such an empirical Padé-type approximant are determined, $g(\theta)$ can be calculated by replacing $I(z)$ in (2.17) by the ansatz function (2.18). One readily obtains

$$g(\theta) = \frac{1}{2\pi} \frac{1 + (d_1 - d_2)n_1 - d_2 + (1 - d_1 + d_2)n_1 \cos\theta - (1 - d_2 + n_1)\cos 2\theta}{1 + d_1^2 + d_2^2 + 2d_1(1 + d_2)\cos\theta + 2d_2 \cos 2\theta}.$$
$$(2.19)$$

Obviously, the above procedure is applicable also in the case of experimental magnetization data. In a first step, the capability of this approach will be checked on numerical data sets of the isothermal magnetization in the case of the $S = 1/2$ mean-field model and the linear Ising chain [27]. Figure 2.3 shows the normalized isothermal magnetization which originates from the mean-field equation [28]

$$\tanh\left(\frac{gS\mu_B\mu_0 H}{k_B T}\right) = \frac{\frac{m}{m_s} - \tanh\left(\frac{mT_c}{m_s T}\right)}{1 - \frac{m}{m_s}\tanh\left(\frac{mT_c}{m_s T}\right)} \qquad (2.20)$$

at $T = 1.15T_c$. The solid line shows the best fit of (2.18) to the data. In addition to n_1, d_1 and d_2 the fitting parameter $a = 2gS\mu_B\mu_0/k_B T$ enters the fitting function by means of $z = \exp(-2gS\mu_B\mu_0 H/k_B T)$. The resulting values of the parameters read $n_1 = -1.54$, $d_1 = -1.256$, $d_2 = 0.347$ and $a = 4.076$. The inset of Fig. 2.3 exhibits the corresponding density function which is determined from (2.19) by substitution of the fitting parameters. It is in good agreement with the results of Kortman and Griffiths [24]. In particular, the square-root-type onset of $g(\theta)$ at $\theta = \theta_g$ is evidenced from the best fit shown in the log-log plot of the density function, which yields the exponent $\mu = 0.52 \approx 1/2$. As a second check of the approach, Fig. 2.4 shows the normalized isothermal magnetization of the linear Ising chain. It is calculated from the free-energy per spin [28]

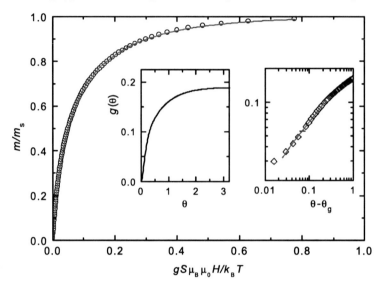

Fig. 2.3. Best fit (*solid line*) of the ansatz function (2.18) to the numerical data of the isothermal magnetization (*circles*) obtained from the mean field (2.20). The corresponding density function $g(\theta)$ is shown on a linear (*left inset*) and log-log scale (*right inset*), respectively. The latter points out the square-root-type onset of $g(\theta)$ at θ_g in accordance with the slope $b \approx 1/2$ of the linear regression (*line*)

$$\frac{F}{N} = -k_B T \ln \left(e^{J/k_B T} \cosh h/k_B T + \sqrt{(e^{2J/k_B T} \sinh^2 h/k_B T + e^{-2J/k_B T})} \right).$$
(2.21)

The resulting magnetization reads [17]

$$\frac{m}{m_s} = \sqrt{\frac{z^2 - 2z + 1}{z^2 - 2z(1 - 2e^{-4J/k_B T}) + 1}}.$$
(2.22)

The best fit of (2.18) to the magnetization data which are calculated from (2.22) with $k_B T/J = 3.5$ yields $n_1 = -1.003, d_1 = -0.394, d_2 = 0.520$ and $a = 1.978$. The inset of Fig. 2.4 exhibits the density functions which originate from (2.19) (solid line) and the rigorous result

$$g(\theta) = \frac{1}{2\pi} \frac{\sin \theta/2}{\sqrt{\sin^2 \theta/2 - e^{-4J/k_B T}}}$$
(2.23)

for $\theta > \theta_g = \arccos(1 - 2e^{-4J/k_B T})$ of Lee and Yang (dashed line), respectively [17]. Roundings of the pole at θ_g reveal the limitations of the simple ansatz function (2.18). They become crucial when critical exponents of the so-called Lee–Yang edge singularities are analyzed. In that case, refinements of the ansatz become necessary. They will be considered in Sect. 2.5. Remarkably, however, although the singularity is not well described by the analytical ansatz (2.18), θ_g is precisely obtained from the fitting procedure when identifying the gap angle with the position of the bending point at $\theta = 1.2$ indicated by the vertical dashed line.

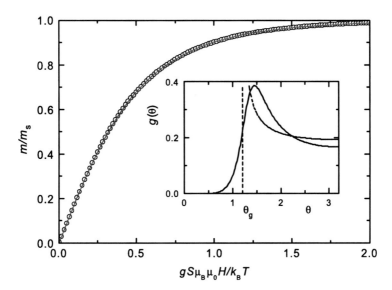

Fig. 2.4. Best fit (*solid line*) of the ansatz function (2.18) to the numerical data of the isothermal magnetization (*circles*) of the linear Ising chain determined from (2.22). The *inset* shows the density functions which originate from (2.19) (*solid line*) and the rigorous result of Lee and Yang [17] (*dashed curve*), respectively. The *vertical dashed line* indicates the singularity of the rigorous density function at $\theta = \theta_g$ and the bending point of $g(\theta)$ originating from the fitting procedure

The above discussion shows that significant differences between the density functions of the linear chain and the mean-field theory emerge from the ansatz function (2.18). This is remarkable in view of the apparent similarity of the corresponding magnetization isotherms (compare Fig. 2.3 and Fig. 2.4). Hence, fitting of isothermal magnetization data with empirical Padé-type functions is a promising approach in order to get access to unknown density functions of various FM Ising model systems.

2.4 Experimental Access to the Density of Lee–Yang Zeros

Anisotropic Heisenberg systems with strong uniaxial anisotropy are the best available approximations of Ising systems. As pointed out in Sect. 2.1 the

Ising-type behavior of $FeCl_2$ originates from its large single-ion anisotropy constant $D/k_B = 9.8$ K. Below the Néel temperature , $T_N = 23.7$ K [29, 13], weak AF inter layer exchange gives rise to three-dimensional order. However, above T_N, strong ferromagnetic correlations arise within the hexagonal Fe^{2+} layers and dominate the weak fluctuations of the 3d AF order-parameter. Hence, $FeCl_2$ becomes a model system for 2d ferromagnets on a triangular lattice.

Figure 2.5 shows the magnetic phase diagram H_a vs. T of $FeCl_2$, where the applied magnetic field H_a is aligned parallel to the c axis. The critical line starts at $T_N = 23.7$ K and separates the paramagnetic (PM) from the AF phase. At the tricritical point a crossover from critical behavior into a first-order spin-flip transition sets in. Sketches of spin configurations of the AF and the PM saturated states are indicated by arrows. A typical domain configuration of AF/PM coexistence is depicted between the upper and the lower boundaries of the mixed region. The domain structure has been visualized by Faraday microscopy [30] at $T = 10$ K and $H_a = 1.03$ MA/m. The width of the meander-type AF (black) and PM (white) domains is about $5 \, \mu m$.

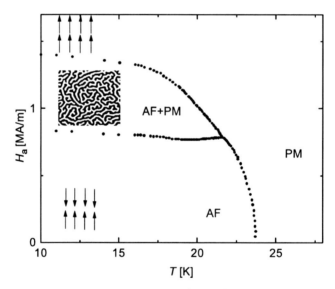

Fig. 2.5. Magnetic phase diagram H_a vs. T (*circles*) of $FeCl_2$. Spin configurations of the AF and PM saturated state are depicted by arrows. A typical configuration of coexisting AF/PM domains which are measured by Faraday microscopy is shown between the *upper* and *lower boundaries* of the mixed region (AF + PM)

Figure 2.6 shows the isothermal magnetization data which are measured by SQUID magnetometry on an as-cleft c platelet with thickness $t = 0.5$ mm and area $A = 18$ mm^2 at temperatures $34 \leq T \leq 99$ K. The data sets are normalized with respect to the saturation moment, $m_s \approx 4$ kAm^2. It is deduced from the high-field limit of the m vs. H data at $T = 4.5$ K. They are shown in Fig. 2.6 by a dashed line. In accordance with the magnetic phase diagram of Fig. 2.5 at this temperature, FeCl$_2$ behaves as a prototypical metamagnet switching from long-range AF into saturated PM order at $H \approx$ 0.8 MA/m. Note that all data have been corrected for demagnetization effects using $H = H_a - NM$. N is the demagnetizing factor, which is calculated according to $N = (dM/dH_a)^{-1} \approx const.$ for $0.8 < H_a < 1.3$ MA/m, within the coexistence region of the AF and PM phases [31].

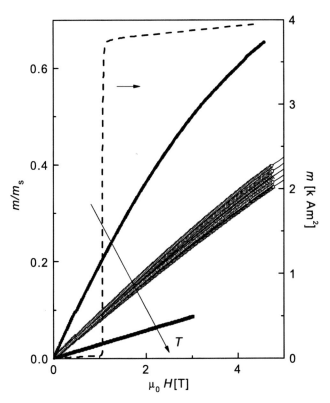

Fig. 2.6. m/m_s vs. H of FeCl$_2$ for temperatures $T = 34, 49, 50, 51, 52, 53$ and 99 K (*solid circles, down triangles, open squares, crosses, open circles, up triangles* and *solid squares*, respectively) and $T = 4.5$ K (*dashed line*). Results of the best fits of (2.18) to the data at $T = 49, ..., 53$ K are indicated by *full lines*

The corrected isothermal magnetic moment m/m_S vs. H of FeCl$_2$ [29, 13] is shown for temperatures $T = 34, 49, 50, 51, 52, 53$ and 99 K, respectively. The data are best-fitted to (2.18). The results of the fitting procedure and its high-field extension is indicated for $T = 34$ K in Fig. 2.7 by a full line

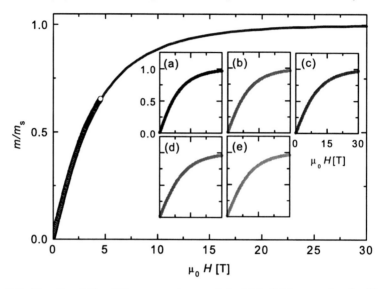

Fig. 2.7. Results of the fitting procedure and its high-field extension for $T = 34$ K (*full line*) together with the magnetization data (*circles*) involved in the fit. The insets (**a**)–(**e**) show the fitting results in the extended magnetic field range for temperatures $T_N = 49\,K + n\Delta T$, where $\Delta T = 1$ K and $n = 0, 1, ..., 4$, respectively

together with the magnetization data (circles) involved in the fit. The insets (a)–(e) show in detail the fitting results in the extended magnetic field range for the temperatures $T_N = 49\,K + n\Delta T$, where $\Delta T = 1$ K and $n = 0, 1, ..., 4$, respectively.

Figure 2.8 shows the density functions, $g(\theta)$, which correspond to the magnetization data obtained at $T = 49, ..., 53$ K (curves 1–5) and $T = 99$ K (curve 6). They have been calculated via (2.19) by inserting the fitting parameters n_1, d_1 and d_2 obtained from (2.18) and summarized in Tale. 2.2. The Landé factor $g = 4.1$ [32] enters z and, hence, (2.18), as a fixed parameter. In accordance with the simple example shown in Fig. 2.2 (square) the zeros of the partition function of a non-interacting system accumulate at $z = -1$, which corresponds to a density function $g(\theta) = \delta(\theta - \pi)$. Hence, the pronounced peak of $g(\theta, T = 99\,K)$ at $\theta = \pi$ (Fig. 2.8, curve 6) reflects the limit of weak interaction, $k_B T \gg J$. With decreasing temperature the maximum value of $g(\theta)$ decreases, while its position shifts towards lower θ values. For example at $T = 51$ K (curve 3), $g(\theta)$ is nearly zero for $0 < \theta < 0.8$, but exhibits a steep increase with increasing θ, which yields a maximum of $dg/d\theta$ at $\theta = 1.5$. The pronounced peak at $\theta = 1.8$ is followed by a smooth decay into the constant value $g(\theta = \pi) = 0.19$. This behavior bears similarity with the θ dependence of the density function of the square lattice , which was theoretically determined at $T = 6T_c$ [24]. The observed steep increase of $g(\theta)$ indicates the upper bound of the gap, $g(\theta) = 0$ for $0 < \theta < \theta_g$, while the pronounced peak in curve 3 reflects the smeared singularity at $\theta = \theta_g$. This smearing originates from the truncation of the Padé-type expansion used in (2.18). This was clearly evidenced in the case of the linear chain, whose exactly known [24] singularity becomes rounded within the Padé-type approach (see above). In addition, however, one has to take into account the dependence of $m(H)$ on the details of the lattice structure. Hence, one might also expect qualitative differences between the underlying density functions of the square and the triangular lattices . Tentatively, this might be at the origin of the discrepancy between the steep decay of $g(\theta \to \pi)$ in the case of the square lattice [24] and its virtual absence in the case of the triangular lattice (Fig. 2.8).

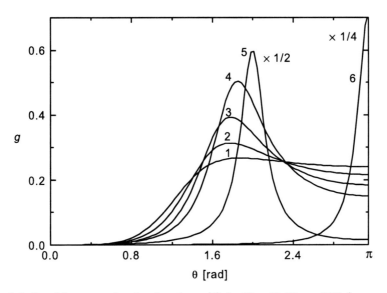

Fig. 2.8. Lee–Yang zero-density function $g(\theta)$ for $T = 49, 50, ..., 53\mathrm{K}$ (curves 1–5, respectively) and $T = 99$ K (curve 6). Data of curve 5 and 6 are scaled by factors $1/2$ and $1/4$, respectively

It is a remarkable property of all density functions which are constructed according to (2.17) that the constraint of (2.12) is fulfilled for any set of parameters provided by the fitting procedure. Moreover, the constraint is even independent of the details of the empirical fitting function. This is on the one hand checked by numerical integration of the density functions of Fig. 2.8 and, on the other hand, easily proved in general on the basis of Cauchy's formula

$$I(z) = \frac{1}{2\pi i} \oint \frac{I(\xi)}{\xi - z} \mathrm{d}\xi. \qquad (2.24)$$

In the limit of infinite magnetic fields, z approaches $\lim_{h \to \infty} z = 0$, while the normalized magnetization saturates at $I(z = 0) = 1$. Hence, substitution of

$z = 0$ into (2.24) yields

$$2\pi i = \oint \frac{I(\xi)}{\xi} d\xi. \tag{2.25}$$

On integrating along the unit circle $\xi = e^{i\theta}$ one obtains with $d\xi = i\xi d\theta$

$$2\pi = \int_0^{2\pi} I(e^{i\theta}) d\theta = \int_0^{2\pi} \mathrm{Re} I(e^{i\theta}) d\theta + i \int_0^{2\pi} \mathrm{Im} I(e^{i\theta}) d\theta, \tag{2.26}$$

where $\int_0^{2\pi} \mathrm{Im} I(e^{i\theta}) d\theta = 0$ has to be fulfilled, because 2π is a real number. The explicit representation of the density function according to (2.17) provides

$$\mathrm{Re}\, I(e^{i\theta}) = 2\pi g(\theta). \tag{2.27}$$

Substitution of (2.27) into (2.26) finally yields the constraint $\int_0^{2\pi} g(\theta) d\theta = 1$.

Table 2.2. Best-fit parameters n_1, d_1, d_2 of (2.18) to the m/m_s vs. H data partially displayed in Fig. 2.6, $J/k_B = -(T/12) \ln((d_1 - n_1)/2)$ and quality parameter χ^2, which measures the sum of the squares of the deviations of (2.18) from the respective data points normalized to the degrees of freedom

$T\,[K]$	n_1	d_1	d_2	$J/k_B\,[K]$	χ^2
34	-1.55830	-1.07621	0.44795	4.031	2.45×10^{-7}
35	-1.62526	-1.09583	0.42893	3.877	4.66×10^{-7}
36	-1.60725	-1.04905	0.41056	3.829	6.26×10^{-7}
49	-1.22375	-0.33542	0.28636	3.314	2.86×10^{-7}
50	-1.00352	-0.12235	0.35496	3.415	1.84×10^{-7}
51	-0.79428	0.07411	0.44338	3.546	1.79×10^{-8}
52	-0.58248	0.28942	0.52122	3.598	6.10×10^{-8}
53	-0.10146	0.71108	0.74829	3.978	2.28×10^{-8}
99	-0.41602	1.50348	0.54979	0.339	5.90×10^{-9}

It is remarkable that the isothermal magnetization data exhibit quite a lot of non-trivial information about the properties of the density functions like, e.g., the systematic temperature dependence of the gap angle, although some of the magnetization isotherms reach less than 40% of the saturation magnetization in the available field range $\mu_0 H \leq 5$ T and, hence, exhibit only weak curvature.

Generally, in order to obtain meaningful fitting results from slightly structured data, it is necessary to minimize the noise level of the measurement. Our SQUID magnetometer (Quantum Design MPMS-5S), being capable of detecting a minimal magnetic moment between 1×10^{-10} and 5×10^{-10} Am^2 within the entire field range and having a temperature stability better than

10 mK fulfills this condition in a nearly ideal manner. Nevertheless, it is necessary to check the validity of the above procedure. The magnetic specific heat, $c(H)$, is a thermodynamic quantity, which can be deduced from $m(H)$ but also from the free-energy expression (2.21) involving the density $g(\theta)$. Both results are shown in Fig. 2.9 by the functions $\Delta c(H)$ and $c(H)$, respectively.

Their comparison is a check of the consistency of the applied method of analysis. The function $\Delta c = (c(H) - c(H = 0))/m_s$ is calculated from the fitting results shown in Fig. 2.7 (insets (a)–(e)) according to

$$\Delta c = T \int_0^H \partial^2(m/m_s)/\partial T^2 \mathrm{d}H'. \tag{2.28}$$

Equation (2.28) is derived from Maxwell's relation [33]. It allows the comparison of experimental magnetization and specific heat data and has been successfully applied in the case of the related metamagnetic compound $FeBr_2$ [4]. The second-order derivative which enters the integral of (2.28) is approximately calculated by [23]

$$\frac{\partial^2 m}{\partial T^2} = \frac{-m(T = 49\,\mathrm{K}) + 16m(50\,\mathrm{K}) - 30m(51\,\mathrm{K}) + 16m(52\,\mathrm{K}) - m(53\,\mathrm{K})}{12\,\mathrm{K}^2}. \tag{2.29}$$

An alternative method to obtain $c(H)$ utilizes the density function $g(\theta)$ in conjunction with the magnetic free-energy function of (2.21). By inserting the $g(\theta)$ functions shown in Fig. 2.8 and the respective $z(T, H)$ values one obtains F vs. H curves, which are displayed in Fig. 2.9 for $T = 49, 50, 51, 52$ and 53 K (curves 1–5, respectively).

From these functions the magnetic entropy, s, and the specific heat, c, are calculated by numerical derivation as shown for $T = 51$ K in Fig. 2.9 (inset) and Fig. 2.10 (open circles), respectively. It is seen that, apart from small deviations which originate from errors of the numerical integration, both curves $\Delta c(H)$ and $c(H)$ yield identical results, where $c(H = 0) = -\Delta c(H \to \infty)$ as expected.

Since the curvature of m/m_s $mathrm, vs.\,H$ is weak in the available field range of the magnetometer, the extrapolation of the fitting results into the high-field regime is risky and has to be checked separately. Therefore, a second independent check is performed. The leading term of the high-field series of $f(z)$, (2.18), is calculated. It reads

$$f(z) = 1 - (d_1 - n_1)\,z + O[z^2]. \tag{2.30}$$

Comparison with the theoretical expansion of the magnetization of the two-dimensional Ising ferromagnet on a triangular lattice [25] yields $u^3 = 1/2\,(d_1 - n_1)$, where u is given by $u = \exp(-4J/k_B T)$. Hence, the ferromagnetic nearest-neighbor in-plane exchange constant of $FeCl_2$ is related to the fitting parameters n_1 and d_1 according to $J/k_B = -(T/12)\ln\left((d_1 - n_1)/2\right)$.

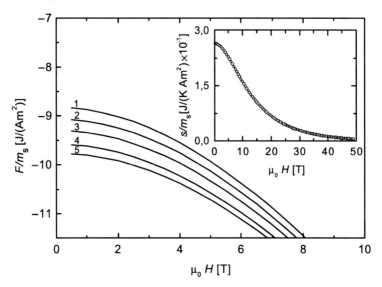

Fig. 2.9. Field dependence of the free-energy calculated from $g(\theta)$ at $T = 49, ..., 53$ K (curves 1–5, respectively). The *inset* shows the field dependence of the entropy at $T = 51$ K calculated from the free-energy curves 1,2,4 and 5 by numerical derivation

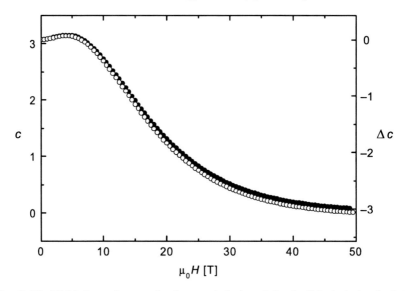

Fig. 2.10. Field dependences of c (*open circles*) and Δc (*solid circles*) calculated from $g(\theta)$ and by use of Maxwell's relation, (2.28), respectively

The average value of the exchange constant yields $J/k_B = 3.7$ K for $34 \leq T \leq 53$ K. Within an error of $\approx 6\%$ this is in accordance with the value $J/k_B = 3.94$ K which has been determined from magnon-dispersion data of inelastic neutron-scattering investigations on $FeCl_2$ [13]. Disappointingly, the value from the data at $T = 99$ K, $J/k_B = 0.339$, deviates from the experimental one by one order of magnitude. Presumably this is due to the finite value of the single-ion anisotropy of $FeCl_2$ [13], which violates the condition of Ising-type symmetry at high temperatures.

2.5 Yang–Lee Edge Singularities

The Padé-type fitting function, (2.18), which has been introduced in Sect.2.3, is an appropriate ansatz in order to elaborate the qualitative θ dependence of the zero-density functions in the entire range $0 \leq \theta \leq \pi$. However, the numerical test of the approach by means of the linear Ising chain has revealed limitations of the analysis. Singularities of the density functions, which indicate critical behavior of the ferromagnet in the presence of purely imaginary magnetic fields, are smeared out into residual finite peaks. In order to study these singularities in detail on the basis of real experimental data, a refinement of the fitting function is required. Moreover, the magnetic field range of the measurement has to be enlarged in order to improve the reliability of the fitting parameters.

The nature of the exotic criticality of the Ising ferromagnet at θ_g is related to its conventional critical behavior at the Curie temperature, T_c, where the temperature-driven phase transition takes place. Above T_c, a gap angle $\theta_g > 0$ separates all complex zeros from the real axis of the z-plane. Hence $g(\theta) = 0$ for $|\theta| < \theta_g$. In accordance with the Padé approach, θ_g depends on temperature and becomes zero at $T = T_c$, where complex zeros touch the real axis. The zero density which evolves at $T < T_c$ determines the spontaneous normalized magnetization $I = m/m_s$ of the Ising ferromagnet according to $I = 2\pi g(0)$.

At the critical temperature the dependence of the magnetization on small magnetic fields is given by $m \propto H^{1/\delta}$ [34]. In the case $\delta > 1$, this scaling behavior generates a singularity of the magnetic susceptibility $\chi = \partial m/\partial H$ at $H = 0$. The non-analytic behavior of the magnetization at $T = T_c$ originates from the accumulation of complex zeros on the real axis at $z(H = 0) = 1$ [18]. In analogy with the non-analytic behavior of m at $T = T_c$ for $H = 0$, non-analyticity also arises at $T > T_c$ for the purely imaginary field $H_0(T) = -i\theta_g k_B T/(2gS\mu_B\mu_0)$, which defines the onset of the zero density $g(\theta) > 0$ at $z(H = H_0) = \exp(i\theta_g)$ [35]. On approaching θ_g from above, the density function exhibits a power law behavior $g \propto (\theta - \theta_g)^\mu$, which is the manifestation of the Yang-Lee edge singularity . In accordance with this singularity at $\theta = \theta_g$, the generalized susceptibility also diverges at $H = H_0$ in the case of $\mu < 1$. Both $\mu < 1$ and the universality of μ are assumed to hold in general [36]. In the case of a 2d Ising ferromagnet, a rigorous theory predicts $\mu = -1/6$ [37].

As pointed out by Kortman and Griffiths [24], the ansatz function

$$m \propto \tau(\tau^2 + \tan^2 \theta_g/2)^\mu \qquad (2.31)$$

with $\tau = (1 - z)/(1 + z)$ implies a density function with an asymptotic behavior of the type $g \propto (\theta - \theta_g)^\mu$ when approaching θ_g from above. It is, hence, a good candidate for an appropriate fitting function. Unfortunately, in the case $\mu > -1/2$ an artificial singularity of the density function at $\theta = \pi$ originates from the divergence of the function $\tau(z)$ at $z = \exp(i\pi) = -1$. Therefore, the function $\tau(z)$ is modified in such a way that the singularity of $g(\theta)$ at $\theta = \pi$ is suppressed. However, the new function $\tilde{\tau}(z)$ still has to conserve both the basic property $\tilde{\tau}(z = 1) = 0$ and the essential condition $\lim_{\theta \to \theta_g} \tilde{\tau} = -i \tan \theta_g/2$, which reveals the physical singularity of $g(\theta)$ at $\theta = \theta_g$. It can be verified, that these conditions are fulfilled most easily by the expression

$$\tilde{\tau} = 1/2(1 - z) \left[1 - \frac{\tan \theta_g/2}{\sin \theta_g}(z - \cos \theta_g) \right]. \qquad (2.32)$$

Substitution of (2.32) into the proportionality (2.31) and normalization of the resulting ansatz function with respect to the saturation magnetization m_s yields

$$I = K\tilde{\tau} \left(\tilde{\tau}^2 + \tan^2(\theta_g/2)\right)^\mu / \left(\tilde{\tau}(0) \left(\tilde{\tau}(0)^2 + \tan^2(\theta_g/2)\right)^\mu\right). \qquad (2.33)$$

In order to take into account small errors which might be involved in the normalization procedure of the data, a proportionality constant $K \approx 1$ is introduced as a fitting parameter in addition to the physically essential parameters μ and θ_g.

Two crucial points have to be checked with respect to the capability of the above approach in order to rely on the analysis of the experimental data. First of all, in the case of exactly solvable models the correct critical exponents μ have to be generated from the analysis of the isothermal magnetization. Secondly, the possible relevance of simpler theoretical concepts, which may also provide satisfactory descriptions of the magnetization isotherms, has to be ruled out.

As an example the linear Ising chain is chosen, which represents an exactly solvable model system with a gap exponent $\mu = -1/2$. Here a mean-field approach yields a seemingly good fit to the exact magnetization data which, however, ends up with the wrong gap exponent $\mu = +1/2$. Figure 2.11a shows the calculated isothermal magnetization (circles) of the linear Ising chain for $k_B T/J = 100$ [17]. In Fig. 2.11c these data are fitted to the mean-field equation (2.20) of an $S = 1/2$ ferromagnet in arbitrary dimension [28].

Although the fit appears quite satisfactory, two deficiencies become apparent at second sight. First, a finite best-fitted value $T_c/T = 0.02$ is clearly at odds with the non-ordering linear chain. Second, as shown elsewhere [24] the critical behavior attributed to (2.20) yields an exponent $\mu = +1/2$ in extreme contrast with the exact one, $\mu = -1/2$ [17]. On the other hand, when fitting the data to (2.31) (Fig. 2.11a, solid line) one readily obtains $\mu = -0.50$ and $\theta_g = 2.74$. Figure 2.11b shows a log–log plot of the density function, which is known from [17]. The linear fit (line) of slope $\mu = -0.5$ indicates the asymptotic critical behavior of the rigorous expression of (2.23).

The above example demonstrates that simplicity of an approach cannot in general be a guideline for the choice of a theoretical description. It is well-known from the analysis of critical behavior, that in particular a mean-field analysis will rarely provide correct critical exponents although it may sometimes fit the data with sufficient accuracy.

In addition to the refinement of the fitting function the analysis of critical behavior requires an improvement of the reliability of the parameters obtained from the fitting procedure. Therefore, the investigated field range has been extended from 5 T [26] to 12 T [27] with respect to the SQUID measurements reported in Sect. 2.4. A vibrating sample magnetometer (VSM, Oxford Instruments MagLab^{VSM}) provides the high magnetic fields in combination with required high resolution. The VSM measurements are carried out on an as-cleft square rectangular c platelet with thickness $t = 0.3$ mm and an area $A = 4$ mm^2. The sample is mounted in a small gelatin capsule. The capsule is filled with cotton wool in order to prevent any movement of the sample. The magnetic moment of the sample holder turns out to be less than 0.2% of the magnetic moment of the sample at $T = 50$ K and $\mu_0 H = 12$ T. Hence, no background correction of the data has to be taken into account within the analysis. The increased curvature of the isotherms in the high-field regime increases the accuracy and stability of the fitting procedure. Hence, the least-squares fit yields unambiguous sets of parameters within the framework of the ansatz function of (2.33).

Figure 2.12 shows the isothermal magnetization for $T = 4, 34, 49, 50$ (inset a), 51, 52 (inset b) and 53 K in internal axial magnetic fields $0 \leq \mu_0 H \leq 12$ T. Note that the plotted number of data points in Fig. 2.12 is reduced by a factor of 50 for $T = 4$ K and by a factor of 10 for the data sets at $34K \leq T \leq 53$ K with respect to the total number of measured and analyzed data points.

The results of the best fits are shown in Fig. 2.12 as solid lines for all isotherms at $34\,K \leq T \leq 53$ K. Moreover, Fig. 2.13 shows the density functions g vs. $\theta - \theta_g$ for $T = 49, 50, 51, 52$ and 53 K, respectively on a log–log scale. They are calculated by using (2.17) after substitution of the fitting parameters μ and θ_g into (2.31).

According to the asymptotic power-law behavior of $g(\theta)$ near to θ_g, all curves show linear behavior with equal slopes, $\mu = -0.15 \pm 0.02$, in good agreement with the theoretical predictions for 2d Ising ferromagnets [36, 37].

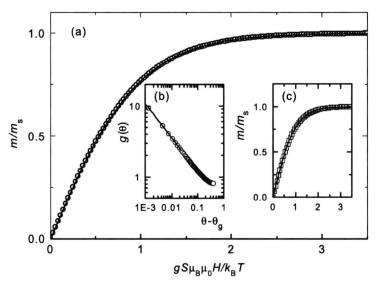

Fig. 2.11. (a) $m/m_\mathrm{s\,vs.}$ $gS\mu_\mathrm{B}\mu_0H/k_\mathrm{B}T$ of the linear Ising chain for $k_\mathrm{B}T/J = 100$. The *line* represents the best fit of (2.31) to the data. *Inset* (b) shows a log–log plot of the rigorous density function [17] with the linear fit of slope $\mu = -1/2$ in the asymptotic regime. *Inset* (c) shows the magnetization data (*squares*) with the result of the best fit of (2.20)

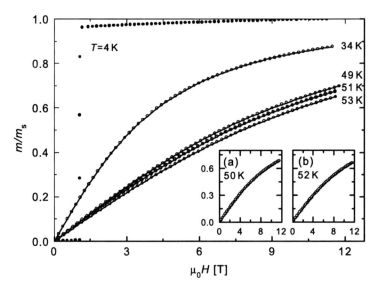

Fig. 2.12. Isothermal magnetization data $m/m_\mathrm{s\,vs.}$ H of FeCl$_2$ measured by VSM magnetometry for temperatures $T = 4, 34, 49, 50$ (inset (**a**)), $51, 52$ (inset (**b**)) and 53 K, respectively. The densities of plotted data points are reduced by a factor of 50 for $T = 4$ K and by a factor 10 for $34\,K \leq T \leq 53$ K. Best fits of (2.33) to the data sets for $T \geq 34$ K are shown by *solid lines*

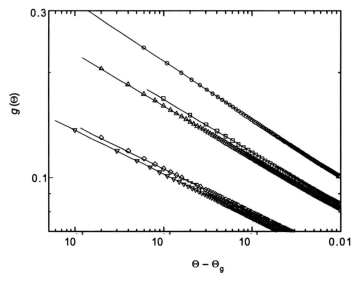

Fig. 2.13. Density functions g vs. $\theta - \theta_g(T)$ for $T = 49$ (*circles*), 50 (*squares*), 51 (*up triangles*), 52 (*down triangles*) and 53K (*diamonds*) on a log–log scale. The slopes of the best-fitted power-law functions (*solid lines*) yield Yang–Lee edge exponents $\mu = -0.15 \pm 0.02$

Moreover, all curves exhibit a gap region of zero density and a pronounced singularity on approaching θ_g from above. The gap angle decreases from $\theta_g = 0.73$ rad at $T = 50$ K towards $\theta_g = 0.63$ rad at $T = 34$ K. This reflects the expected gradual decrease of θ_g with decreasing temperature. The T dependence of θ_g determines the T dependence of $H_0(T)$, which defines the critical line of the Yang–Lee edge singularity. Although these critical lines are not universal, a lot of theoretical work has been done in order to determine details of their behavior for various systems [38, 39].

Despite the reasonable T dependence of θ_g and the 2d Ising criticality of g at θ_g for isotherms at $T \geq 49$ K, there is a strong deviation of the Yang–Lee edge exponent $\mu = -0.365$ for $T = 34$ K. This is attributed to the weak AF superexchange $J'/k_B = -0.18$ K between Fe^{2+} ions on adjacent (111) layers of the rhombohedral $FeCl_2$ crystal, which gives rise to a crossover into 3d long-range AF order. Although the crossover takes place continuously, a rapid increase of the 3d fluctuations is expected when decreasing the temperature from $T = 49$ K to $T = 34$ K. This precursor phenomenon indicates the subsequent power-law divergence of the AF correlations length near T_N.

Further limitations of the ideal 2d Ising behavior of $FeCl_2$ come up in the high temperature limit at $T >> 50$ K, which prevent the determination of a meaningful critical exponent. According to the finite single-ion anisotropy of the $S = 1$ effective-spin system, the Ising-type character of $FeCl_2$ breaks down at high temperatures. Despite the thermal excitation of transverse spin components, the FM exchange energy becomes small in comparison with the thermal energy and the system behaves essentially paramagnetically. In that case all zeros of the ideal paramagnetic Ising system accumulate at $z = -1$, which corresponds to a delta function of the zero density peaking at $\theta = \pi$ [26]. This gap angle quantifies the separation of the singularity from the real positive axis of the complex plane. Hence, one cannot expect to obtain appropriate information about the singularity in the high- temperature limit $\theta_g \approx \pi$ from data which are expected to lie outside the critical region.

Nevertheless, in an intermediate-temperature range the isothermal VSM high-field magnetization data of the prototypical 2d Ising ferromagnet $FeCl_2$ can be used in order to determine the Yang–Lee edge exponent μ. The edge exponent $\mu = -0.15 \pm 0.02$ is in good agreement with the theoretical prediction [36, 37] for all investigated isotherms at $49\,K \leq T \leq 53$ K. Thus, for the first time a way has been pointed out which provides experimental access to the theoretical concept of non-analyticity of the magnetization for $T > T_c(H = 0)$ in a purely imaginary magnetic field, although imaginary fields are, of course, experimentally not realizable.

In order to study the criticality of $g(\theta)$ in closer vicinity to T_c future experiments should focus also on 2d Ising ferromagnets as represented by ultra-thin layers with uniaxial anisotropy [40]. Possible candidates are, e.g., Co monolayers on Cu(111) substrates [41] or Ni(111) layers consisting of less than six monolayers on W(110) [42, 43]. Note that the microscopic details of the uniaxial anisotropy are crucial for the selection of an appropriate model system in order to be sure that the anisotropy is maintained at $T > T_c$. This cannot straightforwardly be deduced from the phenomenological anisotropy constants which are expected to vanish above T_c [44]. Moreover, while investigating m vs. H in the temperature range $T > T_c$ care has to be taken that the magnetic properties of the film–substrate system are not spoilt by thermal interdiffusion [45]. Hence, systems with sufficiently low values of T_c have to be chosen.

2.6 Modern Approaches Beyond the Lee–Yang Theory

2.6.1 Fisher Zeros

The circle theorem has been derived by Lee and Yang for Ising ferromagnets only. In fact, inspection of the elementary example given in Sect. 2.2 shows that the circle theorem does not hold in the case of AF interaction. Nevertheless, it is a challenging task to determine the loci of the zeros in the complex z plane for an AF system. In the case of AF interaction the complex zeros are also distributed on loci which exhibit remarkable symmetries. Their investigation started with the work of [20, 22, 46, 47]. However, it is still far from being complete.

From an experimental point of view systems are required whose high temperature fluctuation spectra are completely dominated by AF correlations. However, the analysis of such experimental data requires much more theoretical input and is therefore far more involved. Up to now, there is no experimental attempt to determine the zero distribution of an AF system in the z plane.

It was Fisher [48] who first stressed the fact that a polynomial representation of the partition function (2.3) also applies in the complex temperature variable

$$x = e^{-\beta J}. \tag{2.34}$$

This pioneering work has been continued by several authors [49, 50, 51, 52, 53]. The new approach exhibits its full capability when AF interaction is involved. Here, for the sake of completeness, the basic ideas are briefly summarized from a simplifying experimentalist's point of view.

In analogy to (2.4) Z can be rearranged into

$$Z = e^{\frac{N}{2} q \beta \frac{J}{2}} \sum_n \tilde{K}_n x^n, \tag{2.35}$$

where q is the coordination number of the lattice [48]. The usefulness of this description becomes obvious when looking at the loci of the zeros of the Ising partition function in the complex v plane where v and x are related according to

$$v = \tanh\left(\beta \frac{J}{2}\right) = \frac{1-x}{1+x}. \tag{2.36}$$

Fisher [48] pointed out that in the case of the 2d Ising system on a square lattice the zeros of $Z(h = 0)$ are located on two circles in the complex v plane. These circles are shown in Fig. 2.14. Their parameter representations read

$$v(\theta) = \pm 1 + \sqrt{2}\, e^{i\theta}, \tag{2.37}$$

where $v_{FM}(\theta) = -1 + \sqrt{2} e^{i\theta}$ represents the circle in the case of FM interaction, while $v_{AF}(\theta) = 1 + \sqrt{2} e^{i\theta}$ determines the AF circle.

From (2.37) one easily finds the intersections between the circles and the real v axis according to

$$Im\left(\pm 1 + \sqrt{2}\, e^{i\theta}\right) = 0. \tag{2.38}$$

Substitution of the solutions $\theta = 0$ and $\theta = \pi$ into Re $v(\theta)$ provides the positions of the branch cuts $v(\theta = 0) = \pm 1 + \sqrt{2}$ and $v(0 - \pi) - \pm 1 - \sqrt{2}$. Two of these four solutions correspond to complex temperatures and are, hence, non-physical results, while $v_{AF}(\theta = \pi) = -(\sqrt{2}-1)$ and $v_{FM}(\theta = 0) = \sqrt{2}-1$ represent the AF and FM transition points in accordance with the rigorous analysis of the 2d Ising system on a square lattice, respectively [54, 55].

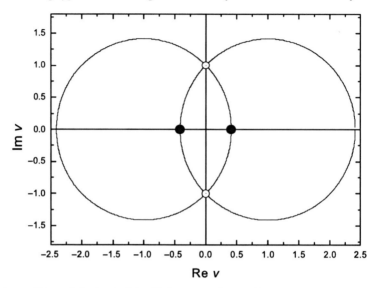

Fig. 2.14. Plots of the loci of the Fisher zeros for a 2d Ising system in the complex v plane. The *full circles* indicate the intersections with the real v axis which belong to the FM and AF transition points, respectively. The *open circles* indicate the zeros of the partition function (2.40) of the toy model (2.5)

Again it turns out to be instructive to have a closer look at the toy model (2.5) of two interacting Ising spins. Its partition function in the case of zero magnetic field reads

$$Z = 2(e^{\beta J} + e^{-\beta J}) \tag{2.39}$$

which can be rearranged into

$$Z = 2e^{\beta J}(1 + x^2) \tag{2.40}$$

in accordance with the general representation (2.35) for $q = 2$ and $N = 2$. The solutions of $Z(x) = 0$ read $x = \pm i$, which yields $v = \pm i$ as predicted by (2.36). The result is independent from temperature. Hence, the zeros of the partition function of two interacting Ising spins never touch the real axis, in accordance with the absence of a phase transition in finite systems.

2.6.2 Zeros of the Partition Function in the Case of First-Order Phase Transitions

In their celebrated paper from 1952 Lee and Yang pointed out their amazement about the simplicity and beauty of the circle theorem by two famous quotations. They introduced their paper with the statement "We were quite surprised, therefore, to find that for a large class of problems of practical interest, the roots behave remarkably well in that they distribute themselves not over the complex plane, but only on a fixed circle". Finally, they concluded with the remark "One cannot escape the feeling that there is a very simple basis underlying the theorem, with much wider application, which still has to be discovered" [17]. It seems that more than four decades later the investigation of the zeros of the partition function in the general case of a continuum system in the vicinity of a first-order phase transition came much closer to this simple basis underlying the Lee–Yang theorem [57, 59, 58].

Here the basic elements of this new approach are briefly discussed. Interestingly, the analysis does not start from the polynomial representation of the partition function but is based on the characteristic features of the probability distribution of the order-parameter conjugate to the external ordering field h. The Ising model (2.2) which shows for $d \geq 2$ a field-driven first-order phase transition at $T < T_{\mathrm{c}}$ is an ideal example in order to illustrate the basic elements of this approach.

Again we start from the partition function

$$Z = \sum_n \mathrm{e}^{-\beta E_n} \tag{2.41}$$

of the canonical ensemble , where E_n are the eigenvalues of the corresponding Hamiltonian. In the case of the Ising Hamiltonian (2.2) this yields (2.3), which has been prepared for the polynomial representation (2.4). Equation (2.41) can be expressed alternatively as

$$Z = \int \mathrm{d}\tilde{E}\, \Omega(\tilde{E})\, \mathrm{e}^{-\beta \tilde{E}}, \tag{2.42}$$

where $\Omega(\tilde{E}) = \sum_n \delta(\tilde{E} - E_n)$ is the density of states. The latter is known to determine, e.g., the partition function of the microcanonical ensemble. If we split the total energy $\tilde{E}$ into the exchange energy $E = -\sum_{i,j} J_{ij} s_i s_j$ and the Zeeman energy $-mh = -h \sum_i s_i$ according to

$$\tilde{E} = E - mh, \tag{2.43}$$

(2.42) can be rearranged into

$$Z = \int \mathrm{d}m\, \mathrm{e}^{\beta m h}\, Z_m, \tag{2.44}$$

where $Z_m = \int \mathrm{d}E\, \mathrm{e}^{-\beta E} w(m, E)$ is determined by the density of states $w(m, E)$ for a given moment m and energy E. It is the behavior of $w(m, E)$ and the corresponding Z_m which allows us to generalize the circle theorem. The qualitative behavior of $w(m, E)$ is known from Monte Carlo simulations on systems of finite size [56].

If w is doubly peaked in the two-phase region the locus of the zeros of the partition function in the complex field plane forms a circle near the positive real axis. In particular, if the transition is symmetric, which means that ω is a symmetric function of m, then at least in the thermodynamic limit the zeros of the partition function lie on the unit circle. Finally, if the distribution turns out to be asymmetric, the zeros form a curve, which is in general not a unit circle [57].

Remarkably, the nature of the distribution function is given by the latent heat which is involved in the first- order transition [57]. If the difference of the specific heat capacities of the two phases is zero, the transition is symmetric and the circle theorem holds.

In order to illustrate the situation in the case of a symmetric transition where the Lee–Yang circle theorem holds, we follow the simple example given in [59]. Therefore one assumes that with increasing system size the symmetric doubly peaked distribution can be approximated by two symmetric δ functions. With respect to the energy it is reasonable to assume that the distribution is sharply peaked around the average value $< E >$. Hence, Z_m becomes

$$Z_m \approx C \left(\delta(m- < m >) + \delta(m+ < m >) \right). \tag{2.45}$$

Substitution of (2.45) into (2.44) yields

$$Z = \frac{Z_0}{2} \left(e^{\beta<m>h} + e^{-\beta<m>h} \right) = Z_0 \cosh(\beta < m > h). \tag{2.46}$$

Replacement of the magnetic moment m by the magnetization $M = m/V$ yields the volume-dependent partition function

$$Z = Z_0 \cosh(\beta < M > V h). \tag{2.47}$$

This partition function allows us to study the field-induced first-order phase transition at $h = 0$ in the thermodynamic limit of infinite sample size [60, 61, 62]. To this end we calculate $M = M(h)$ according to

$$M = -\frac{1}{V} \frac{\partial F}{\partial h}, \tag{2.48}$$

where the free-energy is explicitly given by $F = -k_B T \ln(Z_0 \cosh(\beta < M > V h))$. Figure 2.15 shows the isothermal magnetization at $\beta < M > V = 0.5, 1$ and 5, respectively. Inspection of the isotherms in Fig. 2.15 illustrates the evolution of the discontinuity of $M = M(h)$ which sets in at $h = 0$ on approaching the thermodynamic limit $V \to \infty$. The inset of Fig. 2.15 displays the corresponding free-energy curves F vs. h. Their increasing curvature at $h = 0$ reflects the increasing slope of M vs. h at $h = 0$ with increasing volume. As pointed out above, the Lee–Yang circle theorem holds in the case of a symmetric transition. In fact, it is straightforward to show that the zeros of the partition function (2.46) lie on the unit circle in the complex z plane. With $z = e^{-2\beta h}$ (2.46) is rearranged into

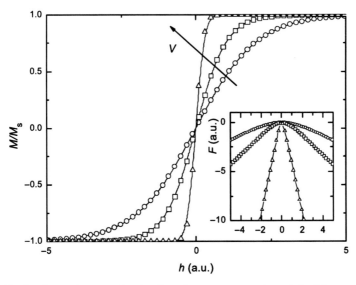

Fig. 2.15. Isothermal magnetization based on a symmetric and doubly peaked distribution for $\beta < M > V = 0.5$ (*circles*), 1 (*squares*)and 5 (*triangles*), respectively. The *inset* shows the corresponding free-energy curves

$$Z = \frac{Z_0}{2}\left(\left(\frac{1}{z}\right)^{\frac{1}{2}<M>V} + z^{\frac{1}{2}<M>V}\right). \tag{2.49}$$

Putting $Z(z) = 0$ yields

$$z^{<M>V} + 1 = 0, \tag{2.50}$$

with its complex solutions

$$z = e^{i(2n+1)\pi/(<M>V)}. \tag{2.51}$$

They lie, indeed, on the unit circle $z = e^{i\theta}$ at $\theta_n = (2n+1)\pi/(< M > V)$ and are separated by the constant angle

$$\Delta\theta = \frac{2\pi}{< M > V} \tag{2.52}$$

as shown in Fig. 2.16.

Although the uniform distribution of the zeros according to $\theta_n = (2n+1)\pi /(< M > V)$ indicates the oversimplification of the model, the essentials of the Yang–Lee theory of phase transitions are completely demonstrated in the framework of the model. In particular, the asymptotic approach of the zeros towards the positive real axis in the thermodynamic limit of infinite volume becomes obvious from (2.52). The gap angle θ_g, which quantifies this approach is directly related to $\Delta\theta$ according to

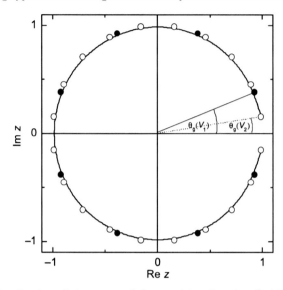

Fig. 2.16. Distribution of the zeros of the partition function (2.46) for two finite volumes V_1 (*solid circles*) and V_2 (*open circles*) with $V_2 = 2.5V_1$. The gap angles $\Delta\theta_g(V_{1,2})$ quantify the distance between θ_0 and the real positive axis for V_1 and V_2, respectively. With increasing system size the gap angle decreases in accordance with the Yang–Lee prediction for a phase transition in the thermodynamic limit

$$\theta_g = \frac{1}{2}\Delta\theta = \frac{\pi}{<M>V}. \tag{2.53}$$

Since the model describes the behavior of the isothermal magnetization at $T < T_c$, $\lim_{V\to\infty}\theta_g(V) = 0$ is in accordance with the Yang–Lee theory of phase transitions [18].

The Yang–Lee theory of phase transitions has been developed in order to understand the origin of the non-analyticity which exhibits the potential of the grand canonical ensemble in the thermodynamic limit of infinite volume on approaching the critical point. The equivalence between the grand canonical partition function of the lattice gas and the partition function of the Ising ferromagnet in the canonical ensemble [17, 48] gives rise to the experimental access to the theory via magnetometry.

Obviously, the rigorous theory of phase transitions is based on pure equilibrium thermodynamics. However, it turns out that the Yang–Lee concept has also a significant impact in the case of non-equilibrium thermodynamics and, in addition, in the case of finite systems. A few important examples should be briefly mentioned here.

It turns out that concepts of equilibrium can be applied to various non-equilibrium systems. Among them driven diffusive systems play an outstanding role. The latter is similar in importance which Ising models possess in the framework of thermal equilibrium. In particular, a grand canonical partition function can be introduced for this non-equilibrium model system [63]. In close analogy with the Yang–Lee theory, the distribution of the zeros of the partition function in the complex fugacity plane can be studied [64]. Again, a first-order phase transition is defined when the zeros touch the real positive axis. In that case, the generalized pressure becomes non-analytic while the density becomes a discontinuous function of the fugacity [64].

In the framework of equilibrium thermodynamics it is obvious that a phase transition requires the limit of infinite volume or particle number. In the case of finite systems the partition function (2.41) is built up from a finite sum of Boltzmann factors $e^{-\beta E_n}$. The logarithm of such a sum has no possibility to show non-analyticity. However, it is meaningful to investigate the precursor behavior of a finite system which undergoes a real phase transition in the limit of infinite particles. Such systems became very popular in recent years. For example, Bose–Einstein condensates were investigated where the condensate originates from a finite number of trapped atoms. For such systems classification schemes have been derived which originate from the distribution of the complex zeros of the partition function. In this framework it makes sense to distinguish between first- and second-order phase transitions where the Ehrenfest criterion is fulfilled in the limit of infinite system size [65].

A similar situation is given when Monte Carlo simulations are used on a finite system size in order to extrapolate to the behavior of infinite systems. In this case, the investigation of critical behavior is traditionally based on finite-size scaling methods. Recently, however, it turned out that the analysis of the density of the zeros of the partition function can be used as a measure of the strength of the phase transition [66].

References

1. H. Eschrig: *The Fundamentals of Density Functional Theory* (Teubner, Leipzig 1996)
2. M. Pajda, J. Kudrnovský, I. Turek, V. Drchal and P. Bruno: Phys. Rev. Lett. **85**, 5424 (2000)
3. M.M.P. de Azevedo, Ch. Binek, W. Kleemann and D. Bertrand: J. Magn. Magn. Mater **140**, 1557 (1995)
4. O. Petracic, Ch. Binek, W. Kleemann, U. Neuhausen and H. Lueken: Phys. Rev. B **57**, R11051 (1998)
5. Ch. Binek, T. Kato, W. Kleemann, O.Petracic, F. Bourdarot, P. Burlet, H. Aruga Katori, K. Katsumata, K. Prokes and S. Welzel: Euro. Phys. J. B **15**, 35 (2000)
6. H. Aruga Katori, K. Katsumata and M. Katori: Phys. Rev. B **54**, R9620 (1996)
7. W. Selke: Z. Phys. B **101**, 145 (1996)
8. M. Pleimling, W. Selke: Phys. Rev. B **56**, 8855 (1997)
9. K. Held, M. Ulmke, N. Blümer and D. Vollhardt: Phys. Rev. B **56**, 14 469 (1997)
10. M. Acharyya, U. Nowak and K.D. Usadel: Phys. Rev. B **61**, 464 (2000)

11. L.J. De Jongh and A.R. Miedema: Adv. Phys. **23**, 1 (1974)
12. Landolt Börnstein: *Non-Metallic Inorganic Compounds Based on Transition Elements, New Ser. III, 27, Subvol. j1* (Springer, Berlin 1994)
13. R.J. Birgeneau, W.B. Yelon, E. Cohen and J. Makovsky: Phys. Rev. B **5**, 2607 (1972)
14. M. Tinkham: *Group Theory and Quantum Mechanics* (McGraw-Hill, New York 1964)
15. U. Balucani and A. Stasch: Phys. Rev. B **32**, 182 (1985)
16. J.C. Slater: *Quantum Theory of Molecules and Solids, Vol. 2* (McGraw-Hill, New York 1965) p. 77 ff
17. T.D. Lee and C.N. Yang: Phys. Rev. **87**, 410 (1952)
18. C.N. Yang and T.D. Lee: Phys. Rev. **87**, 404 (1952)
19. S. Ono, Y. Karaki, M. Suzuki and C. Kawabata: J. Phys. Soc. Japan **25**, 54 (1968)
20. M. Suzuki, C. Kawabata, S. Ono, Y. Karaki and M. Ikeda: J. Phys. Soc. Jpn. **29**, 837 (1970)
21. C. Kawabata and M. Suzuki: J. Phys. Soc. Jpn. **28**, 16 (1970)
22. S. Katsura, Y. Abe and M. Yamamoto: J. Phys. Soc. Jpn. **30**, 347 (1971)
23. I.N. Bronstein and K.A. Semendjajew: *Taschenbuch der Mathematik, 24. Aufl.* (Deutsch, Frankfurt/Main 1989)
24. P.J. Kortman and R.B. Griffiths: Phys. Rev. Lett. **27**, 1439 (1971)
25. M.F. Sykes, J.W. Essam and D.S. Gaunt: J. Math. Phys. **6**, 283 (1965)
26. Ch. Binek: Phys. Rev. Lett. **81**, 5644 (1998)
27. Ch. Binek, W. Kleemann and H. Aruga Katori: J. Phys.: Condens. Matter **13**, L811 (2001)
28. H.E. Stanley: *Introduction to Phase Transitions and Critical Phenomena* (Oxford University Press, New York 1971)
29. L.J. De Jongh and A.R. Miedema: Adv. Phys. **23**, 1 (1974)
30. J. Kushauer: *Magnetische Domänen in verdünnten uniaxialen Antiferromagneten.* PhD thesis, Gerhard-Mercator-Universität Duisburg (1995)
31. J.F. Dillon, E.Y. Chen and H.J. Guggenheim: Phys. Rev. B **18**, 377 (1978)
32. D. Bertrand, F. Bensamka, A.R. Fert, J. Gelard, J.P. Redoulès and S. Legrand: J. Phys. C **17**, 1725 (1984)
33. H. Mitamura, T. Sakakibara, G. Kido and T. Goto: J. Phys. Soc. Jpn. **64**, 3459 (1995)
34. M.E. Fisher: Rep. Prog. Phys. **30**, 615 (1967)
35. J.E. Kirkham and D.J. Wallace: J. Phys. A **12**, L47 (1979)
36. M.E. Fisher: Phys. Rev. Lett. **40**, 1610 (1978)
37. S.N. Lai and M.E. Fisher: J. Chem. Phys. **103**, 8144 (1995)
38. Xian-Zhi Wang and Jai Sam Kim: Phys. Rev. E **57**, 5013 (1998)
39. Xian-Zhi Wang and Jai Sam Kim: Phys. Rev E **58**, 4174 (1998)
40. F.J. Himpsel, J.E. Ortega, G.J. Mankey and R.F. Willis: Adv. Phys. **47**, 511 (1998)
41. J. Kohlhepp, H.J. Elmers, S. Cordes and U. Gradmann: Phys. Rev. B **45**, 12 287 (1992)
42. Y. Li and K. Baberschke: Phys. Rev. Lett. **68**, 1208 (1992)
43. M. Farle: Rep. Prog. Phys. **61**, 755 (1998)
44. A. Hucht and K.D. Usadel: Phys. Rev. B **55**, 5579 (1997)
45. K. Baberschke: Appl. Phys. A **62**, 417 (1996)

46. S.Y. Kim and R.J. Creswick: Phys. Rev. E **63**, 066107 (2001)
47. R.G. Ghulghazaryan, N.S. Ananikian and P.M.A. Sloot: Phys. Rev. E **66**, 046110 (2002)
48. M.E. Fisher: In *Lectures in Theoretical Physics, Vol VIIC*, ed. by W.E. Brittin (Gordon and Breach, New York 1964)p. 1
49. J. Stephenson and R. Couzens: Physica A **129**, 201 (1984)
50. H.J. Brascamp and H. Kunz: J. Math. Phys. **15**, 65 (1974)
51. C.N. Chen, C.K. Hu and F.Y. Wu: Phys. Rev. Lett. **76**, 169 (1996)
52. S.Y. Kim and R.J. Creswick: Phys. Rev. E **58**, 7006 (1998)
53. W.T. Lu and F.Y. Wu: J. Statist. Phys. **102**, 953 (2001)
54. H.A. Kramers and G.H. Wannier: Phys. Rev. **60**, 252 (1941)
55. L. Onsager: Phys. Rev. **65**, 117 (1944)
56. K.C. Lee: J. Phys. A **23**, 2087 (1990)
57. K.C. Lee: Phys. Rev. E **53**, 6558 (1996)
58. J. Lee and K.C. Lee: Phys. Rev. E **62**,4558 (2000)
59. K.C. Lee: Singularities in the Complex Temperature Plane at the First Order Phase Transitions and Critical Points. In: *Statistical Physics, Third Tohwa University International Conference*, ed. by M. Tokuyama and H.E. Stanley (American Institute of Physics, Melville New York 1999) pp. 314–325
60. V. Privman and M.E. Fisher: J. Statist. Phys. **33**, 385 (1983)
61. J. Lee and J.M. Kosterlitz: Phys. Rev. B **43**, 3265 (1991)
62. C. Borgs and R. Kotecky: Phys. Rev. Lett. **68**, 1734 (1992)
63. B. Derrida, S.A. Janowsky, J.L. Lebowitz and E.R. Speer: J. Statist. Phys. **73**, 813 (1993)
64. P.F. Arndt: Phys. Rev. Lett. **84**, 814 (2000)
65. P. Borrmann, O. Müller and J. Harting: Phys. Rev. Lett. **84**, 3511 (2000)
66. W. Janke and R. Kenna: Analysis of the Density of Partition Function Zeroes– A Measure for Phase Transition Strength. In: *Computer Simulation Studies in Condensed-Matter Physics, Vol. XIV*, ed. by D.P. Landau, S.P. Lewis and H.-B. Schüttler (Springer, Berlin 2002) p. 97

Part II

Exchange-Bias

3 Ferromagnetic Thin Films for Perpendicular and Planar Exchange-bias Systems

Perpendicular exchange-bias systems require the growth of ferromagnetic thin films with perpendicular anisotropy on top of uniaxial anisotropic antiferromagnets. After a brief introduction of the phenomenological description of magnetic anisotropy the origin of perpendicular anisotropy in Co/Pt multilayers, their growth and characterization are described.

3.1 Anisotropy

The magnetic anisotropy of a magnetized sample is quantified by the energy which is required in order to rotate the saturation magnetization from its easy towards its hard axis. It is of the order 10^{-6} to 10^{-3} eV per atom. An estimate of the total magnetic energy from the thermal energy $k_B T_C$ at the magnetic transition yields 0.1 eV per atom. Hence, the anisotropy energy causes only a small correction with respect to the total magnetic energy and, therefore, one could argue that it might be of minor relevance. However, most applications, e.g., starting with the Chinese compass, which is the earliest known application of magnetism in history, largely depend on the existence of magnetic anisotropy. Although various contributions to the magnetic anisotropy are discussed in the literature, each of them involves either dipole–dipole interaction or spin–orbit coupling. These mechanisms couple the spin degrees of freedom with the anisotropic lattice and therefore break the rotational invariance of the Hamiltonian.

The dipolar interaction energy W_{ij} between two magnetic moments m_i and m_j with mutual distance r_{ij} reads [1]

$$W_{ij} = \frac{1}{4\pi\mu_o} \left(\frac{m_i \, m_j}{r_{ij}^3} - 3 \frac{(m_i \, r_{ij})(m_j \, r_{ij})}{r_{ij}^5} \right). \qquad (3.1)$$

Its inherent anisotropic properties are exhibited by the explicit r_{ij} dependence. Equation (3.1) reflects the potential energy which the moment m_i gains in the dipolar field of m_j. In a macroscopic body, the total field contribution of all moments which are localized outside the Lorentz sphere that surrounds the moment m_i is summed up by a continuous dipolar field $H_d(r)$.

For a given continuous magnetization distribution $M(r)$ the dipolar field can be calculated according to

$$H_d(r) = -\nabla \Phi(r), \tag{3.2}$$

where in the case of vanishing macroscopic current density the scalar potential [2]

$$\Phi(r) = -\frac{1}{4\pi} \int_V d^3 r' \frac{\nabla M(r')}{|r - r'|} + \frac{1}{4\pi} \int_{\partial V} d^2 r' \frac{M(r')}{|r - r'|} \tag{3.3}$$

is derived from Maxwell's equations . For a homogeneously magnetized body of ellipsoidal shape only the surface term of (3.3) contributes to the magnetic potential. Its gradient yields the uniform demagnetizing field

$$H_d = -\underline{\underline{N}} M, \tag{3.4}$$

where $\underline{\underline{N}}$ is the demagnetizing tensor of trace 1. Thin magnetic films with their normal vectors parallel to the z axis represent the limit of plates of infinite lateral extension. This condition yields the simple demagnetizing tensor

$$\underline{\underline{N}} = \begin{pmatrix} 0\,0\,0 \\ 0\,0\,0 \\ 0\,0\,1 \end{pmatrix}. \tag{3.5}$$

The density of the shape-anisotropy energy is in general given by the potential self-energy which experiences the moment of a magnetized body in its demagnetizing field. It reads

$$F_{shape} = -\frac{\mu_0}{2V} \int_V d^3 r M(r) H_d(r). \tag{3.6}$$

Hence, in the case of homogeneously magnetized thin magnetic films the anisotropy energy reads

$$F_{shape} = \frac{\mu_0}{2} M_z^2 = -\frac{\mu_0}{2} M^2 \sin^2 \theta + const., \tag{3.7}$$

where θ is the angle between the z-axis and the direction of M. Obviously, $\theta = \pi/2$ minimizes the shape anisotropy. Hence, in-plane orientation of the magnetization is favored. Therefore, perpendicular anisotropy of thin magnetic films requires additional crystalline contributions.

The magnetocrystalline anisotropy arises mainly from spin–orbit interaction, but may also involve dipole–dipole interaction. An example for the latter case is given by the antiferromagnet MnF_2, where dipolar interaction is the main source of anisotropy [3, 4]. The analysis of the dipolar crystalline anisotropy is based on a discrete summation of the dipole contributions of the magnetic moments within the Lorentz sphere.

The volume contribution of the magnetocrystalline anisotropy is an intrinsic magnetic property which depends on the symmetry of the crystal. In general, the magnetic free-energy has to be invariant under inversion of the magnetization in accordance with time-inversion symmetry. Hence, the expansion of the free-energy with respect to the direction of the magnetization, M/M, is an even function of the involved angles. Crystalline symmetry reduces the number of independent expansion coefficients. For example, in a system with hcp structure the leading terms of the magnetocrystalline energy density read

$$F_{\text{cryst.}} = K_0 + K_1 \sin^2 \theta + K_2 \sin^4 \theta + K_3 \sin^4 \theta \cos 4\phi + ..., \tag{3.8}$$

where θ denotes the angle between the magnetization and the c axis while ϕ is the azimuth in a spherical coordinate system.

However, in magnetic thin films and multilayers, surfaces and interfaces break the translational symmetry of the bulk systems, which gives rise to anisotropy contributions in addition to (3.8). Atoms at the interface are exposed to electronically modified environments. For example, the strong perpendicular anisotropy of Co/Pt multilayers originates from additional interface-induced anisotropy. A single Co layer of hcp (0001) texture exhibits in-plane anisotropy. However, the magnetic susceptibility $\chi = S\chi_0$ of the adjacent Pt layers shows a strong Stoner amplification of $S = 2$ with respect to the simple Pauli susceptibility χ_0 of a free-electron gas. χ_0 depends only on the electronic density of states, $n(E_F)$, at the Fermi energy E_F. The Stoner enhancement of Pt, however, originates from the exchange splitting, I, of the spin-resolved density of states. It enters the amplification factor S according to $S = 1/(1 - In(E_F))$. Additional strong spin–orbit coupling of Pt drives the total moment perpendicular to the magnetic layer and gives rise to perpendicular anisotropy in the case of appropriate growth conditions.

In addition to the interface-induced break down of the translational invariance, a stress induced elastic deformation of the lattice can also reduce the crystalline symmetry, e.g., from cubic to tetragonal. In that case, the magnetocrystalline anisotropy which corresponds to the reduced symmetry is activated via spin–orbit interaction. In general, magnetoelastic interaction couples the elastic properties of a body to the magnetic degrees of freedom. On the other hand, reciprocity of the magnetoelastic effect gives rise to magnetostriction where a change of the magnetization creates an elastic deformation. Magnetoelastic interaction plays a major role in thin magnetic films. In particular, the growth of heterolayers gives rise to strong elastic deformations as a consequence of lattice mismatches involved in the growth process.

A phenomenological description of the interface anisotropy per unit area is given by [5]

$$f_{\text{int}} = -k_\perp \cos^2 \theta + k_{||} \sin^2 \theta \cos^2 \phi. \tag{3.9}$$

The corresponding anisotropy energy per unit volume reads

$$F_{int} = \frac{2f_{int}}{d}, \tag{3.10}$$

where d is the layer thickness, while the factor 2 takes into account the number of components which build up the interface. In the case of perpendicular n-fold rotation axes with $n > 2$ the in-plane constant $k_{||}$ is zero by symmetry. Hence, the interface anisotropy of hexagonal layers like, e.g., hcp (0001)-oriented Co films simply reads

$$F_{int} = \frac{2}{d}k_\perp \sin^2 \theta + const. \tag{3.11}$$

Summing up the anisotropy contributions pointed out by (3.7), (3.8) and (3.11) one obtains an expression which describes the important case of uniaxial anisotropy and holds, e.g., in the case of hexagonal symmetry. Neglecting higher order anisotropy terms of the magnetocrystalline contribution, (3.8), it reads

$$F = K_{eff} \sin^2 \theta, \tag{3.12}$$

where the effective anisotropy constant K_{eff} is given by

$$K_{eff} = -\frac{\mu_0}{2}M^2 + K_1 + \frac{2}{d}k_\perp. \tag{3.13}$$

In the case $K_{eff} > 0$ the anisotropy energy is minimized at $\theta = 0$ describing perpendicular anisotropy, while $K_{eff} < 0$ yields in-plane anisotropy in accordance with the minimization of F at $\theta = \pi/2$.

Figure 3.1 shows a typical hysteresis loop of the multilayer glass/Pt 50 Å/ $(Co\,3.5\,Å\,Pt\,12Å)_{15}$ /Pt 8Å. The sample was grown at $T = 500$ K under UHV condition by thermal (Co) and electron-beam (Pt) evaporation. The hysteresis was measured by Faraday rotation performed at room temperature using a photoelastic modulation technique with a light source of wavelength $\lambda = 670$ nm [6]. The rectangular shape of the loop indicates the strong perpendicular anisotropy of the multilayer. The corresponding microscopic images (1–6) utilizing the polar Kerr effect illustrate the process of magnetization-reversal. Heterogeneous nucleation sets in at the switching field, and domains with considerable wall roughness grow rapidly with increasing magnetic field (images 1–3). However, close to saturation local demagnetizing fields stabilize the unswitched domains thus giving rise to rounding of the loop [7] (images 4–6). Very similar behavior is well-known from the literature where, e.g., perpendicular anisotropy with $K_{eff} = 865$ kJ/m^3 has been observed for glass/$(Co\,4.5\,Å\,Pt\,17.7\,Å)_{22}$ multilayers [8]. Moreover, $Co_{1-x}Pt_x$ alloys also show perpendicular anisotropy for Pt contents $0.45 < x < 0.9$ [9]. The phenomenological expansion coefficients of (3.8) have a priori no physical meaning. It is, however, possible to determine the coefficients and their temperature dependence from a microscopic Hamiltonian when calculating the free-energy and expanding the resulting expression into a series of the type

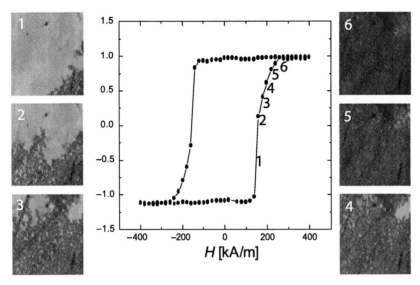

Fig. 3.1. Faraday-rotation measurement (*circles*) of the magnetic hysteresis of glass/Pt 50 Å/(Co 3.5 Å Pt 12 Å)₁₅/Pt 8 Åat T = 300 K using a photoelastic modulation technique with a light source of wavelength λ = 670 nm [6]. Images 1–6 illustrate the magnetization-reversal process by polar Kerr microscopy resolving a lateral region of 120 × 90 mm^2

(3.8) [10]. In particular, such an analysis points out the temperature dependence of the anisotropy constants and the possibility of a temperature induced reorientation transition from, e.g., perpendicular to planar spin alignment becomes obvious.

Section 4.2.2 below discusses the implications of perpendicular anisotropy in the case of heterostructures with unidirectional anisotropy. The latter anisotropy, which has not been discussed so far, has a new quality which arises from exchange coupling at the interface between AF and FM material. This coupling gives rise to the phenomenon of exchange bias.

3.2 Thin-film Growth
on Antiferromagnetic Single Crystals

In our experiments the FM components of the exchange-bias heterosystems have been fabricated from metallic thin films or multilayers. The necessity of thin-film preparation originates from the fact that exchange bias is an interface effect. The metallic films are grown under ultra-high vacuum (UHV) conditions in a multi-chamber system which provides a base pressure of approximately 10^{-10} mbar. Fe, Co and Ag, which possess (at normal pressure) melting temperatures of T_{m} = 1809, 1768 and 1234 K, are deposited by thermal evaporation, while Pt with $T_m = 2045$ K is deposited by electron-beam evaporation onto the surfaces of the AF single crystals $Fe_{1-x}Zn_xF_2$ and $FeCl_2$, respectively [11]. Deposition rates of typically 0.5 Å/s are controlled by piezoelectric quartz resonators during the growth process.

The AF substrates are thermally stabilized at 500 and 425 K, respectively during the deposition of $(Co\,3.5\,Å/Pt\,12\,Å)_n$ multilayers and thin Fe layers of 14 and 5 nm thickness. While the Co/Ptmultilayers are capped with an additional 8 nm Pt layer, Ag layers of 35 nm thickness prevent the thin Fe films from oxidization and allow for ex situ characterization. Ag is used in order to minimize deterioration of the Fe surface by interdiffusion. This is ensured from the very low solubility of Fe and Ag according to the corresponding binary phase diagram [12].

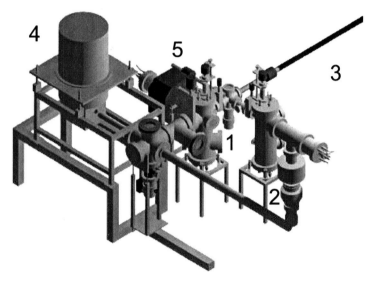

Fig. 3.2. UHV system containing the metal-evaporation chamber (1), the insulator-evaporation chamber (2), a transfer system (3), Kerr magnetometry with an 8 T cryomagnet (4) and the UHV pump system (5)

Figure 3.2 exhibits a detailed drawing of our non-commercial UHV-system [13]. Its basic elements are two separate evaporation chambers for metal growth (1) on the one hand and insulator evaporation (2) on the other hand. The spatial separation of the two evaporation chambers prevents the metal films from contamination with residues of other growth processes. A transfer system (3) allows the transport of the samples, e.g., from the evaporation chambers to the Kerr-magnetometer (4), the low energy electron diffraction (LEED) device and the Auger electron spectrometer (AES), which will be used for structural and chemical analysis, respectively. Ex situ structural analysis is done by small- and large-angle X-ray diffraction (Philips PW 1730) (see Figs. 4.7 and 4.19) and atomic force microscopy (AFM) (Topometrix Explorer) for real space surface analysis. The pump system is based on components which are common in the UHV technique. Turbomolecular pumps (Pfeiffer TMU 065 and TMU 261) provide HV conditions better than 10^{-7} mbar for the metal and the Kerr-rotation chamber, respectively, while the insulator chamber is pumped with an oil diffusion pump (Edwards Diffstack 100). Polyphenyl ether (Santovac 5) is used in order to minimize the remaining vapor pressure of the oil. The UHV condition is reached after baking the whole system to above $T = 420$ K for more than 24 h. After removing the water, the oil diffusion pump and the turbo pump are assisted by a continuously running ion getter pump (Thermionics PS 350, Fig. 3.2 (5)) and a periodically activated titanium sublimation pump (CVT TSPC-1), which provide the final base pressure. The metallic thin films are deposited on $Fe_{1-x}Zn_xF_2$ and $FeCl_2$ single crystals, respectively, in order to grow the model-type AF/FM heterostructures. While the $Fe_{1-x}Zn_xF_2$ crystals are polished to optical flatness with 0.3 mm diamond paste before transferring into the chamber, the $FeCl_2$ crystals are cleft in situ perpendicularly to the c axis (compare Fig. 2.1) in order to get clean (111) surfaces. Figures 3.3 and 3.4 show the 3d AFM images of the topography (a) and the contrast enhanced image of the corresponding force gradient (b) together with a height profile z vs. y (d) recorded along the line $x = 6$ mm and $x = 4.8$ mm of the polished FeF_2 (001) surface and after coverage with a Pt 15 Å/(Co 3.5 Å/Pt 12 Å)$_5$/Pt 8 Åmultilayer [14].

Inspection of the line profile (Fig. 3.3 d) exhibits the effective roughness $R_{ms} = \sqrt{1/N \sum_{i=1}^{N} (z_i - \bar{z})^2}$ of the (001) surface after polishing. On a local scale $0 < \Delta y < 0.3$ μm the roughness is about $R_{ms} \approx 0.8$ nm. A full line-scan $0 < \Delta y < 10$ μm contains contributions of deep and rare scratches which give rise to $R_{ms} \approx 4.8$ nm, a corrugation of nearly 15 lattice constants. Hence, the AFM investigations exhibit rough surfaces of the polished AF insulating substrates. The corresponding interface roughness of the AF/FM heterosystems has a strong influence on the exchange-bias [15, 16, 17, 18]. In particular, the freezing field dependence of the exchange bias field is significantly influenced by the roughness [19]. Details are presented in Sect. 4.2.2 for the perpendicular exchange-bias system $FeF_2(001)/CoPt$.

As expected, after deposition of the metallic multilayer, the corrugation of the surface is drastically reduced. This is suggested after inspection of the topography (Fig. 3.4 a and c) as well as the image of the force gradient (b) and is quantified by the line-scan analysis presented in Fig. 3.4c. On a local scale the effective roughness is reduced from 0.8 to ≈ 0.5 nm, while the full line-scan yields $R_{ms} \approx 2$ nm.

The situation changes significantly for the $FeCl_2$ (111) cleavage planes. Figure 3.5 shows images of the force gradient (a) and the topography (b) of the $FeCl_2$ (111) surface on a large scale of 50 μm $\times$ 50 μm, respectively [6]. The height profile z vs. x along the line $y = 35$ μm (horizontal line in

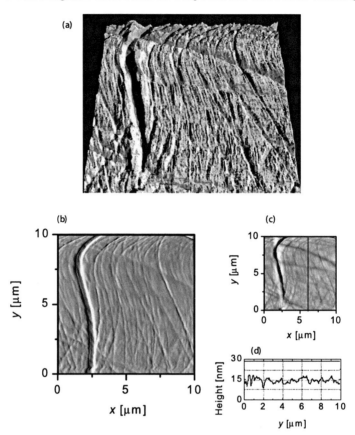

Fig. 3.3. Three AFM images of the FeF_2 (001) surface after polishing to optical flatness. 3d representation of the topography (**a**) and the corresponding image of the force gradient (**b**) are shown together with a height profile z vs. y (**d**) along the line $x = 6$ mm, which is visualized (*line*) in the gray-scale-encoded image of the topography (**c**) (0- to 30 nm height is mapped on a scale between black and white)

Fig. 3.5b) is presented in Fig. 3.5c. Although there is a strong corrugation which gives rise to the roughness $R_{ms} = 158$ nm corresponding to ≈ 90 lattice constants (see Fig. 2.1), there are large terraces of remarkable flatness. Figure 3.6 shows the AFM images which correspond to a region of $5\,\mu m \times 5\,\mu m$ which is indicated in the large scale image of Fig. 3.5a (square). The analysis of the height profile z vs. x along the line $y = 3.2\,\mu m$ (horizontal line in Fig. 3.6b) quantifies the flatness. The roughness along the full line of $5\,\mu m$ length is given by $R_{ms} = 0.8$ nm which is less than 46% of the length of the c axis. Hence, as one may expect, the cleavage planes provide large terraces of high flatness in comparison with the polishing procedure.

The flatness of the terraces on the one hand and the huge roughness on a mesoscopic scale on the other hand favor a strong AF domain formation of

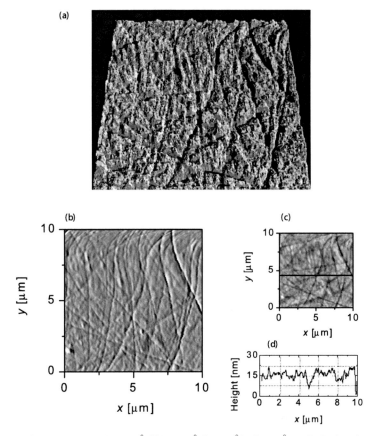

Fig. 3.4. AFM images of Pt 15 Å/(Co 3.5 Å/Pt 12 Å)$_5$/Pt 8 Å on FeF$_2$ (001). 3d representation of the topography (**a**) and the corresponding image of the force gradient (**b**) are shown together with a height profile z vs. x (**d**) along the line $x = 4.8$ mm, which is visualized (*line*) in the gray-scale-encoded image of the topography (**c**) (0- to 30 nm height is mapped on a scale between black and white)

the perpendicular anisotropic FeCl$_2$(111)/CoPt heterosystem. Its magnetic properties are described in detail in Sect. 4.2.3 below.

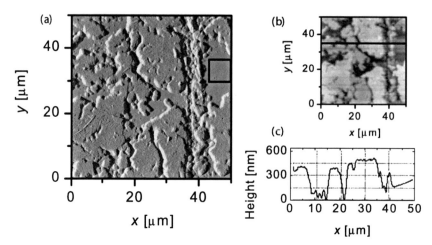

Fig. 3.5. Large-scale AFM images of the $FeCl_2(111)$ cleavage plane. (**a**) and (**b**) show the image of the force gradient and the topography, respectively. The height profile z vs. x along the line $y = 35\,\mu m$ (*horizontal line* in (**b**)) is presented in (**c**). The *square* shown in (**a**) marks the region investigated in detail in Fig. 3.6

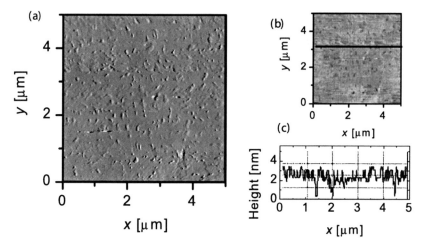

Fig. 3.6. AFM images of the $FeCl_2(111)$ cleavage plane on a magnified scale. (**a**) and (**b**) show the image of the force gradient and the topography, respectively. The height profile z vs. x along the line $y = 3.2\,\mu m$ (*horizontal line* in (**b**)) is presented in (**c**)

References

1. K.H. Hellwege: *Einführung in die Festkörperphysik* (Springer, Berlin 1981) p. 215 ff
2. M.H. Nayfeh and M.K. Brussel: *Electricity and Magnetism* (Wiley, New York 1985) p. 291 ff
3. Y. Shapira: Phys. Rev. B **2**, 2725 (1970)
4. F. Keffer: Phys. Rev. **87**, 608 (1952)
5. B. Hillebrands: *24th IFF-Ferienkurs Magnetismus von Festkörpern und Grenzflächen* (Forschungszentrum Jülich, Institut für Festkörperforschung 1993)
6. B. Kagerer: Senkrechte Anisotropie und exchange bias magnetischer Heteroschichten. Diploma thesis, Gerhard-Mercator-Universität Duisburg (1999)
7. U. Nowak, J. Heimel, T. Kleinefeld, D. Weller: Phys. Rev. B **56**, 8143 (1997)
8. W.B. Zeper, F.J.M. Greidanus, P.F. Carcia and C.R. Fincher: J. Appl. Phys. **65**, 4971 (1989)
9. D. Weller, H. Brändle, G. Gorman, C.J. Lin and H. Notarys: Appl. Phys. Lett. **61**, 2726 (1992)
10. A. Hucht and K.D. Usadel: Phys. Rev. B **55**, 5579 (1997)
11. R.C. Weast and M.J. Astle: *CRC Handbook of Chemistry and Physics*, 63rd edn. (CRC, Boca Raton 1982)
12. T.B. Massalski: *Binary Alloy Phase Diagrams*, Vol.1, 2nd edn. (ASM International, Ohio 1990) p. 35
13. M. Aderholz: *Private communication*, Gerhard-Mercator-Universität Duisburg (2001)
14. S. Kainz: Exchange Bias in senkrecht anisotropen Heteroschichtsystemen. Diploma thesis, Gerhard-Mercator-Universität Duisburg (2000)
15. J. Nogués, D. Lederman, T.J. Moran and I.K. Schuller: Appl. Phys. Lett. **68**, 3186 (1996)
16. A. Mougin, T. Mewes, M. Jung, D. Engel, A. Ehresmann, H. Schmoranzer, J. Fassbender and B. Hillebrands: Phys. Rev. B **63**, 060409 (2001)
17. T. Mewes, R. Lopusnik, J. Fassbender, B. Hillbrands, M. Jung, D. Engel, A. Ehresmann and H. Schmoranzer: Appl. Phys. Lett. **76**, 1057 (2000)
18. Joo-Von Kim, R.L. Stamps, B.V. McGrath and R.E. Camley: Phys. Rev. B **61**, 8888 (2000)
19. B. Kagerer, Ch. Binek and W. Kleemann: J. Magn. Magn. Mater. **217**, 139 (2000)

4 Exchange Bias in Magnetic Heterosystems

This chapter deals with the exchange-bias phenomenon in prototypical magnetic heterostructures. Special emphasis is laid on model systems involving uniaxial anisotropy. Here, the limited number of spin degrees of freedom minimizes the complexity of the problem. The discussion of the exchange-bias effect starts with the presentation of a generalized phenomenological Meiklejohn–Bean approach. It takes into account finite anisotropy and thickness of the antiferromagnetic pinning layer. In addition, microscopic mechanisms which create the antiferromagnetic interface magnetization are investigated and discussed. They are of major importance for the understanding of the exchange-bias effect. This holds, in particular, in the case of antiferromagnetic pinning layers with compensated surfaces. This monograph pays special attention to piezomagnetism and its contribution to the interface magnetization. Piezomagnetism is a well-known bulk phenomenon among the effects of weak ferromagnetism. However, its significance in the framework of the exchange-bias effect has been overlooked so far. In addition, the temperature dependence of the exchange -bias field is studied. The rare case of non-zero exchange bias above the Néel temperature of the pinning layer is presented and analyzed in terms of local quasi-critical temperatures. This analysis takes advantage of the close analogy between the enhancement of the blocking temperature and the non-analytic behavior in the Griffiths phase of dilute magnets. Finally, the training of the exchange-bias effect is investigated. It is a clear signature of the non-equilibrium nature of the exchange-bias phenomenon.

4.1 A Generalized Meiklejohn–Bean Approach

The exchange bias describes a magnetic coupling phenomenon between FM and AF materials. Although this proximity effect implies a mutual interaction between the FM and AF constituents, its most striking feature affects the FM hysteresis which shifts along the magnetic field axis after field-cooling of the heterosystem to below the Néel temperature. In addition to this spectacular effect, the coupling also modifies the coercive field of the ferromagnet. Typically, the coercivity is enhanced and its temperature dependence is related to the temperature dependence of the exchange-bias field that quantifies the shift of the hysteresis [1]. Exchange biasing and coercivity enhancement usually vanish at the blocking temperature in the vicinity of the Néel temperature. Moreover, torque measurements reveal a unidirectional anisotropy which originates from the coupling and is at the heart of the phenomenon [2].

Since the pioneering observation in 1956 of the exchange-bias effect on small ferromagnetic Co particles which are embedded in their AF oxide [3, 4], there is a renewed interest in the investigation of the exchange-bias effect in well-defined FM/AF layered heterosystems. As a typical example Fig. 4.1 exhibits the magnetic hysteresis loop of $Fe_{0.6}Zn_{0.4}F_2(110)/Fe$ 14 nm/Ag 35 nm which generates an exchange bias field of $\mu_0 H_e = -3.1$ mT. While the details of this heterostructure are discussed in paragraph 4.3 the sketches of the FM/AF interfacial spin alignments illustrate the basic influence of the coupling on the magnetization-reversal. On cooling the system in an applied magnetic field to below the Néel temperature, the interface moments of the ordered antiferromagnet couple to the polarized FM interface moments via the exchange interaction J (see Fig. 4.1 a). For simplicity the cartoon-like sketches refer to an uncompensated AF interface and suggest a positive exchange coupling. In contrast with the magnetic hysteresis of a coherently rotating single FM layer, the coupling between the FM layer and the AF substrate pins the ferromagnet. This, on the one hand, hampers the reversal of the magnetization with decreasing magnetic field (Fig. 4.1 a → b → c) but, on the other hand, supports the reversal from negative to positive magnetization with increasing magnetic field (Fig. 4.1 c → d → a). Consequently, the hysteresis loop is shifted by $\mu_0 H_e$ along the field axis.

Beyond this intuitive picture, a more quantitative description of the coupling has been introduced by Meiklejohn and Bean (MB) [4, 5]. They started from the well-established Stoner–Wohlfarth free-energy expression which describes the coherent hysteretic magnetization-reversal processes of single-domain particles and magnetic thin films [6]. In order to take into account the interaction between the FM/AF interface moments they added a bilinear exchange which gives rise to an additional unidirectional anisotropy energy. Under the assumption of a perfect FM/AF interface there is in general no quantitative agreement between the MB model and experimental results. Extrapolating from the exchange energies in the bulk material to the strength of the coupling at the interface, the MB approach overestimates the exchange-bias effect by typically two orders of magnitude in comparison with experimental findings [7, 8, 9, 10]. Although the actual microscopic interface interaction is in general unknown, this extrapolation is usually judged as a failure of the MB model which has stimulated an abundance of theoretical and experimental work.

One of the rare, albeit also indirect attempts to determine the modified interlayer and intralayer exchange coupling has been done in the case of Eu monolayers deposited on Gd(0001) [11]. The layer-resolved magnetization has

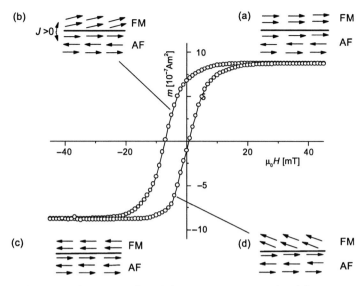

Fig. 4.1. Magnetic hysteresis (*circles*) of a $Fe_{0.6}Zn_{0.4}F_2(110)/Fe$ 14 nm/Ag 35 nm heterostructure measured by SQUID magnetometry after saturation of the FM layer and subsequent field-cooling at $\mu_0 H = 5$ mT to $T = 10$ K. Sketches a–d illustrate the magnetization-reversal of the pinned FM layer in the framework of a Meiklejohn–Bean model. *Arrows* indicated the FM and AF magnetic moments, respectively. *Horizontal lines* represent the FM/AF interface where FM exchange $J > 0$ gives rise to the exchange-bias effect

been measured by X-ray magnetic circular dichroism in photoemission and calculated in the framework of a layer-dependent mean-field approach, respectively. In this particular case, comparison between the experimental and theoretical results reveals in fact a FM interlayer coupling which is close to the exchange constant of Gd bulk material, but gives rise to a noncollinear spin structure in the Eu layer. An additional failure of the MB ansatz originates from the neglect of interface roughness and domain formation. Both idealizations are typically not justified in real heterostructures. As a rare exception, epitaxially grown multilayer systems of artificially structured antiferromagnets and ferromagnets reveal perfect agreement between the MB model and the experimental results [12]. This basic confirmation as well as the simplicity of the MB model make it, however, a favorable first approach in order to interpret experimental data. In view of this simplicity it is surprising that most of the experimental facts are at least qualitatively described within the framework of the MB approach.

In the limit of infinite anisotropy of the antiferromagnet, the MB model yields the simple, but powerful formula [4]

$$\mu_0 H_e = -\frac{J S_{AF} S_{FM}}{M_{FM} t_{FM}}. \tag{4.1}$$

It exhibits the well-known dependences of $\mu_0 H_e$ on the FM layer thickness t_{FM} [13], on the magnetization of the FM layer [14] and on the interface magnetizations S_{AF} and S_{FM}. No information about the origin of S_{AF} and S_{FM} is provided by the MB model. In addition, the coupling constant J enters the MB approach only as a phenomenological constant in order to overcome the problems described above. A lot of theoretical work, which tackles the microscopic foundation of these parameters or develops alternative descriptions of the undirectional anisotropy, has been done [15]. Nevertheless, already the simple MB formula points out the necessity of net magnetic moments at the interface, not only on the FM, but also on the AF side in order to obtain finite exchange-bias. Without exception this holds also in the case of so-called compensated AF surfaces, where $S_{\mathrm{AF}} \neq 0$ requires deviations from the ideal AF order [16].

Although the MB model possesses only a phenomenological character, (4.1) does not express its full capability. It turns out that a generalization of the model allows, for instance, the description of the dependence of the exchange-bias field on the AF layer thickness, t_{AF}. This is, e.g., observed in experiments on NiFe/FeMn heterostructures [17] and in Monte Carlo studies on Ising systems of FM and diluted AF layers [18]. Figures 4.2 and 4.3 exhibit the experimental and simulational results, respectively.

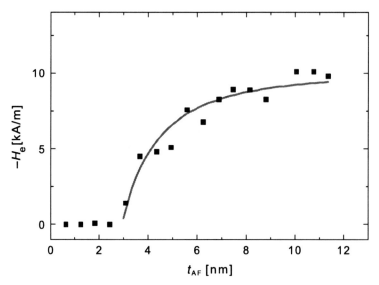

Fig. 4.2. Thickness dependence, $H_e \, vs. \, t_{\mathrm{AF}}$, of the exchange-bias field in $\mathrm{Ni_{80}Fe_{20}/FeMn}$ [17]. The *line* shows the best fit of (4.8) to the data at $t_{\mathrm{AF}} > 3$ nm

The subsequent analysis [19] reveals that both the non-vanishing bulk AF magnetization M_{AF} and a finite angle θ between the applied magnetic field

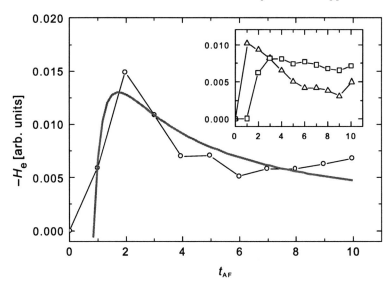

Fig. 4.3. Data from Monte Carlo simulations [18] of the exchange-bias field as a function of the number of AF monolayers, t_{AF}, are shown for 40% (*circles*), 30% (*triangles* in the *inset*) and 60% (*squares* in the *inset*) quenched site dilution of the AF. The *bold solid line* is a best fit of (4.10) to the data at 40% dilution

and the magnetic easy axes tune the details of the t_{AF} dependences. It is the aim of this analysis to exploit the full capability of the MB approach in analytic expressions and overcome restrictions due to partial solutions accounting either for θ [7, 20] or t_{AF} [2] alone.

The starting point of the analysis is the free-energy per unit area [21] completed by adding a Zeeman term involving M_{AF},

$$F = -\mu_0 H M_{FM} t_{FM} \cos(\theta - \beta) - \mu_0 H M_{AF} t_{AF} \cos(\theta - \alpha)$$
$$+ K_{FM} t_{FM} \sin^2 \beta + K_{AF} t_{AF} \sin^2 \alpha - J S_{AF} S_{FM} \cos(\beta - \alpha). \quad (4.2)$$

Here H is the applied magnetic field and $M_{FM/AF}$, $t_{FM/AF}$, $K_{FM/AF}$ and $S_{FM/AF}$ are the absolute values of the magnetization, the layer thickness, the uniaxial anisotropy constant and the interface magnetizations of the FM/AF layer. The latter can be interpreted as macroscopic moments because the MB model assumes parallel orientation of all FM moments during the entire process of coherent rotation. Hence, the FM spins fulfill the condition $s_i^{FM} = s^{FM} \forall i$, and the interaction of the microscopic spins at the interface can be transformed into an interaction of the macroscopic interface moments according to $\sum_{i,j} s_i^{FM} s_j^{AF} \propto S_{FM} S_{AF}$. They are coupled via J, the exchange interaction constant. θ, β and α are the angles between H, M_{FM} and M_{AF} and the FM/AF anisotropy axis, which are parallel aligned for simplicity (see Fig. 4.4). The AF bulk magnetization, M_{AF}, is usually assumed to be zero. This is reasonable in the case where the sublattice magnetizations mutually compensate in the long-range AF ordered state. However, this is no longer the case in diamagnetically diluted AF systems. They are known to decay into a random-field-induced domain state with frozen excess magnetization when cooling to below T_N in an external magnetic field [23]. This mechanism is at least one important possibility to control the appearance of interface magnetization, $S_{AF} \neq 0$, at compensated AF surfaces and thus enables exchange-bias [16, 18]. On the other hand, the excess bulk magnetization, $M_{AF} \neq 0$, of a quenched AF domain state may also be important by virtue of the corresponding Zeeman energy term in (4.2). Surprisingly, metastable domain states can also be induced in non-diluted AF pinning layers. They are probably due to interface roughness [24, 25] and give rise to both M_{AF} and excess susceptibility.

In the case of infinite anisotropy K_{AF}, the minimization of the free-energy yields $\alpha = 0$. Hence, in the case of strong but finite anisotropy, a series expansion of (4.2) with respect to $\alpha = 0$ is reasonable. It reads

$$
\begin{aligned}
F \approx & - JS_{AF}S_{FM}\cos\beta - \mu_0 HM_{FM}t_{FM}\cos(\theta - \beta) \\
& - \mu_0 HM_{AF}t_{AF}\cos\theta + K_{FM}t_{FM}\sin^2\beta \\
& + \alpha\left(-JS_{AF}S_{FM}\sin\beta - \mu_0 HM_{AF}t_{AF}\sin\theta\right) \\
& + \alpha^2\left(K_{AF}t_{AF} + \frac{1}{2}JS_{AF}S_{FM}\cos\beta + \frac{1}{2}\mu_0 HM_{AF}t_{AF}\cos\theta\right). \quad (4.3)
\end{aligned}
$$

This expression is minimized with respect to β and α, which is physically equivalent to the determination of the equilibrium angles β_{eq} and α_{eq} of vanishing torque.

$\partial F/\partial\alpha = 0$ yields

$$
\alpha_{eq} = \frac{JS_{AF}S_{FM}\sin\beta + \mu_0 HM_{AF}t_{AF}\sin\theta}{2K_{AF}t_{AF} + JS_{AF}S_{FM}\cos\beta + \mu_0 HM_{AF}t_{AF}\cos\theta}. \quad (4.4)
$$

From $\partial F/\partial\beta = 0$ we determine the magnetic fields H_{c1} and H_{c2}. They fulfill the condition $M_H(H_{c1}) = M_H(H_{c2}) = 0$, where $M_H = M_{FM}\cos(\theta - \beta)$ is the magnetization component of M_{FM} pointing parallel to the applied magnetic field (see Fig. 4.4).

M_H is the experimentally relevant FM magnetization component which is measured by standard scalar magnetometry. In order to obtain explicit expressions for H_{c1} and H_{c2} we insert $\alpha = \alpha_{eq}$, $\beta_1(M_H = 0) = \theta - \pi/2$ and $\beta_2(M_H = 0) = \theta - 3\pi/2$ into $\partial F/\partial\beta = 0$. Expansion of $\partial F/\partial\beta$ to first-order with respect to $M_{AF} \approx 0$ yields two corresponding linear equations in H, which provide H_{c1} and H_{c2}, respectively. The exchange-bias field is then calculated according to

$$
H_e = (H_{c1} + H_{c2})/2. \quad (4.5)
$$

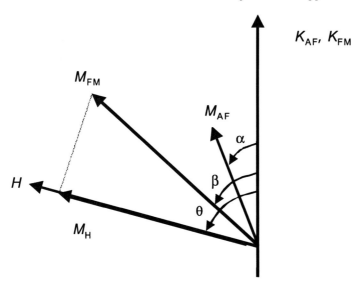

Fig. 4.4. Vector diagram showing the angles α, β and θ related to the orientation of the net AF magnetization M_{AF}, the magnetization of the ferromagnet M_{FM} and the applied magnetic field H with respect to the easy axis of the antiferro- and ferromagnet designated by the corresponding anisotropy constants K_{AF} and K_{FM}, respectively. M_H indicates the projection of M_{FM} onto the field direction

Although the calculation is straightforward, the result is lengthy. In order to simplify the resulting expression, H_e is again expanded into a Taylor series with respect to $M_{\mathrm{AF}} \approx 0$ and $1/K_{\mathrm{AF}} \approx 0$ up to first and second order, respectively. The approximation of strong anisotropy, $1/K_{\mathrm{AF}} \approx 0$, is consistent with the series expansion, (4.3). One finally obtains

$$
\begin{aligned}
\mu_0 H_e = {} & - \frac{J S_{\mathrm{AF}} S_{\mathrm{FM}} \cos\theta}{M_{\mathrm{FM}} t_{\mathrm{FM}}} \\
& - \bigg(\frac{J S_{\mathrm{AF}} S_{\mathrm{FM}} \cos\theta}{16 K_{\mathrm{AF}}^2 M_{\mathrm{FM}}^2 t_{\mathrm{AF}}^2 t_{\mathrm{FM}}^2} \ \big(-4 J K_{\mathrm{AF}} M_{\mathrm{AF}} S_{\mathrm{AF}} S_{\mathrm{FM}} t_{\mathrm{AF}}^2 \\
& + J^2 M_{\mathrm{FM}} S_{\mathrm{AF}}^2 S_{\mathrm{FM}}^2 t_{\mathrm{FM}} + J K_{\mathrm{FM}} M_{\mathrm{AF}} S_{\mathrm{AF}} S_{\mathrm{FM}} t_{\mathrm{AF}} t_{\mathrm{FM}} \\
& - J S_{\mathrm{AF}} S_{\mathrm{FM}} \ (-4 J K_{\mathrm{AF}} M_{\mathrm{AF}} t_{\mathrm{AF}}^2 \\
& + 3 J M_{\mathrm{FM}} S_{\mathrm{AF}} S_{\mathrm{FM}} t_{\mathrm{FM}} + 4 K_{\mathrm{FM}} M_{\mathrm{AF}} t_{\mathrm{AF}} t_{\mathrm{FM}}) \cos 2\theta \\
& + 3 J K_{\mathrm{FM}} M_{\mathrm{AF}} S_{\mathrm{AF}} S_{\mathrm{FM}} t_{\mathrm{AF}} t_{\mathrm{FM}} \cos 4\theta \bigg) .
\end{aligned}
\tag{4.6}
$$

In the limit of infinite anisotropy K_{AF}, (4.6) yields the θ-dependent expression

$$
\mu_0 H_e = - \frac{J S_{\mathrm{AF}} S_{\mathrm{FM}} \cos\theta}{M_{\mathrm{FM}} t_{\mathrm{FM}}},
\tag{4.7}
$$

which was derived previously [7]. In particular, (4.7) provides, again, the basic MB expression in the case $\theta = 0$, which implies parallel orientation of the applied field with the easy axes.

The simplest possible t_{AF} dependence is derived from (4.6) in the limit $M_{AF} = 0$ and $\theta = 0$ and finite, but strong anisotropy K_{AF}. It reads

$$\mu_0 H_e = -\frac{JS_{AF}S_{FM}}{M_{FM}t_{FM}} + \frac{J^3 S_{AF}^3 S_{FM}^3}{8K_{AF}^2 M_{FM}t_{FM}t_{AF}^2}. \tag{4.8}$$

Equation (4.8) qualitatively explains the steep increase and the subsequent saturation of $|\mu_0 H_e|$ with increasing AF layer thickness. Such behavior is not described within the alternative random-field approach of Malozemoff [9], but has been reported by various authors [17, 21, 26]. While Xi and White [26] introduced a more complicated ansatz involving a helical structure of the AF magnetic moments, it is the aim of the present analysis to stress the capabilities of the MB model.

The existence of a critical AF layer thickness was already pointed out by MB. It can simply be motivated from (4.4) on applying the condition $\alpha_{eq} \ll \pi$, as required within the range of validity of (4.3). In the limit $M_{AF} = 0$ (4.4) yields

$$\alpha_{eq} = \frac{\sin\beta}{(2K_{AF}t_{AF})/(JS_{AF}S_{FM}) + \cos\beta}. \tag{4.9}$$

In order to prevent the unphysical divergence of α_{eq} for any direction of M_{FM}, $0 \le \beta \le \pi$, the condition $|K_{AF}t_{AF}| > |JS_{AF}S_{FM}/2|$ has to be fulfilled and, hence, the existence of a critical AF layer thickness becomes obvious. Substitution of the conventional expression of the critical AF layer thickness [4] $t_{AF}^{cr} = |JS_{AF}S_{FM}/K_{AF}|$ yields $|\alpha_{eq}| < 1/\sqrt{3}$ for $0 \le \beta \le \pi$ which proves that in fact $\alpha_{eq} \ll \pi$ is fulfilled for $t_{AF} > t_{AF}^{cr}$.

Figure 4.2 shows the best fit of (4.8) to the $\mu_0 H_e$ vs. t_{AF} data [17] of a $Ni_{80}Fe_{20}$ layer with thickness $t_{FM} = 6.5$ nm deposited on top of FeMn for $3\,\text{nm} < t_{AF} < 12$ nm. The two-parameter fit yields $JS_{AF}S_{FM}/(M_{FM}t_{FM}) = 0.013$ T and $J^3 S_{AF}^3 S_{FM}^3/(8K_{AF}^2 M_{FM}t_{FM}) = 1.08\,10^{-19}$ T/m^2. With $t_{FM} = 6.5$ nm and $M_{FM}(Ni_{80}Fe_{20}) = 0.73$ MA/m [27] we obtain the coupling energy $|JS_{AF}S_{FM}| = 6\,10^{-5}$ J/m^2 and the AF anisotropy $K_{AF} = 7.3\,10^3$ J/m^3. The latter is of the same order of magnitude as the K_{AF}-values obtained e.g. by Mathieu et al. [28] from Brillouin light scattering investigations and by Parkin and Speriosu [2] from torque measurements. The above expression of the critical thickness then yields $t_{AF}^{cr} = 8$ nm, which lies, however, 2.7-times above the steep increase of $\mu_0 H_e$ vs. t_{AF} shown in Fig. 4.2. Apparently, the situation can be improved when setting $t_{AF}^{cr} = JS_{AF}S_{FM}/2K_{AF}$ which emerges as the lower bound of t_{AF} values, fulfilling the condition $|K_{AF}t_{AF}| > |JS_{AF}S_{FM}/2|$. In that case, the remaining error is reduced to less than 34%. Note, however, that the inequality is only a necessary condition. It is not obvious that its lower boundary can be identified with the critical thickness. This numerical discrepancy may originate from the strong simplifications which underlay (4.8). One of them, the assumption $M_{AF} = 0$, will be discussed below.

Besides this numerical inconsistency (see above), the simple $1/t_{AF}^2$ dependence of (4.8) does not model any kind of a peak-like structure in the $|\mu_0 H_e|$ vs. t_{AF} dependence. This is, however, known from experiments, e.g., on $Ni_{80}Fe_{20}/FeMn$ [17] or Fe_3O_4/CoO [29] bilayers as well as from Monte Carlo simulations [18]. In accordance with these findings, (4.6) exhibits the possibility of a competing $1/t_{AF}$ term in the case of $M_{AF} > 0$ and $\theta \neq 0$. The latter condition is not obvious, however, closer inspection of (4.6) shows that the $1/t_{AF}$- terms cancel each other in the case $\theta = 0$. This is in full agreement with results from Monte Carlo simulations on heterostructures of diluted antiferromagnets and FM layers [18]. As discussed above, the diluted antiferromagnet breaks into a random-field domain state on cooling to below the Néel temperature in the presence of an applied magnetic field. These random-field domains carry a frozen net magnetization M_{AF}. Within the framework of the MB approach, (4.6) opens the possibility for a t_{AF} dependence of the type

$$\mu_0 H_e = a + \frac{b}{t_{AF}} + \frac{c}{t_{AF}^2}. \tag{4.10}$$

Figure 4.3 shows the result of a best fit of (4.10) to the Monte Carlo data of [18] which are obtained on a heterosystem with 40% of quenched dilution of the AF-sites (see [18] for details). The peak-structure of $|\mu_0 H_e|$ vs. t_{AF} is qualitatively reproduced with fitting parameters $a = -9 \times 10^{-4}$, $b = -0.042$ and $c = 0.037$ involving units adapted to the Monte Carlo data. Moreover, the simulations show that the peak strongly decreases for 30% as well as 60% dilution. This is reasonable, because the maximum frozen AF moment is expected to be induced by the maximum random-field. In accordance with the approximation $h_r \propto \sqrt{x(1 - x)}$, the magnitude of the random-field, h_r, maximizes at 50% dilution, $x = 0.5$ [30]. Obviously, both concentrations reduce the magnetic moment of the AF layer with respect to the case of 40% dilution. In agreement with the MB approach, (4.6), a reduction of M_{AF} gives rise to a reduced peak height. Hence, in accordance with the results of the Monte Carlo simulations the analysis suggests that, apart from the random-field enhanced AF interface magnetic moment S_{AF}, the magnetization of the subsequent AF layers strongly influences the $\mu_0 H_e$ vs. t_{AF} behavior. While M_{AF} is a well-known feature of field-cooled dilute antiferromagnets, it seems to occur quite generally also in pure AF pinning substrates. One of the alternative sources of M_{AF} is provided by piezomagnetism . Its significance for the exchange-bias is discussed in Sect. 4.3.

4.2 Perpendicular Anisotropic Heterostructures with Uncompensated Interfaces

4.2.1 Influence of the Pinning Layer Susceptibility

As pointed out in the last section, the simplicity of the MB approach originates from the assumption of a coherently rotating ferromagnet which couples via a perfectly flat interface to an antiferromagnet of infinite uniaxial anisotropy. Serious complications arise when leaving this ideal situation. In particular, the ratio of the anisotropies of the AF and FM layers strongly influences the exchange-bias effect. Comparison of the basic MB formula (4.1) with the generalized expression of (4.6) indicates the increasing complexity of the exchange-bias effect already in the case of strong but finite AF anisotropy.

From a phenomenological point of view it is obvious that a huge AF susceptibility favors deviations from the long-range-ordered AF structure which evolves on cooling the antiferromagnet under the perturbing influences of the adjacent FM layer and the applied magnetic freezing field. A magnetized AF state represents a strong deviation from the AF ground state. In the case of a high AF susceptibility the presence of the exchange and the applied magnetic field induce a magnetic moment which can be partly frozen-in on cooling the system to below T_N. As a result, a metastable AF domain state evolves.

$FeCl_2$ for instance exhibits a huge parallel zero-field susceptibility at the Néel temperature $T_N = 23.7$ K in comparison with, e.g., FeF_2. This difference subdivides the systems into two classes in the following denoted as 'soft' and 'strong' antiferromagnets. On the one hand, the uncompensated (001) surface of the strong antiferromagnet FeF_2 pins a perpendicular anisotropic FM top layer while the influence of the ferromagnet on the AF pinning layer is negligible. On the other hand, in heterostructures based on the soft antiferromagnet $FeCl_2$, the AF long-range order breaks into a metastable domain state on field-cooling. This magnetic roughness originates from the topological roughness of the uncompensated (111)-interface layer of the antiferromagnet which couples to the homogeneously polarized ferromagnet via exchange interaction. Hence, in the case of soft AF pinning layers, the mutual interaction at the interface has a strong disordering influence on the antiferromagnet. It strongly suppresses the exchange-bias because the AF domain state is highly susceptible to axial magnetic fields which rearrange the spin alignment during the hysteresis loop. Moreover, the AF domains exhibit a thermally activated relaxation towards their long-range-ordered ground state. Hence, an effective pinning of the FM layer is hampered in the case of soft antiferromagnets. The details of the temperature- and field-dependent susceptibility of such a heterosystem will be discussed in Sect. 4.2.3.

Figure 4.5 shows the temperature dependence, χ' vs. T, of the real part of the magnetic low-frequency susceptibility of $FeCl_2$ and $Fe_{0.47}Zn_{0.53}F_2$, respectively. Inspection reveals that the susceptibility of $FeCl_2$ at $T_N = 23.7$ K (circles) is about 8 times larger than the susceptibility of $Fe_{0.47}Zn_{0.53}F_2$

at $T_N = 36.4$ K (squares). The diamagnetic dilution of FeF_2 shifts the Néel temperature from $T_N(x = 0) = 78$ K down to $T_N(x = 0.53) = 36.4$ K which is close to the Néel temperature of $FeCl_2$ and makes the comparison of the two systems easier.

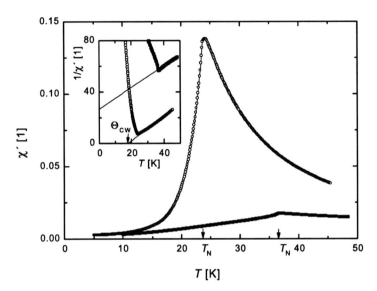

Fig. 4.5. Temperature dependence, χ' vs. T, of the real part of the magnetic low-frequency susceptibility of $FeCl_2$ (*circles*) and $Fe_{0.47}Zn_{0.53}F_2$ (*squares*), respectively. *Arrows* indicate the Néel temperatures at $T_N = 23.7$ and 36.4 K, respectively. The *inset* shows the corresponding $1/\chi'$ vs. T dependence of $FeCl_2$ and $Fe_{0.47}Zn_{0.53}F_2$, respectively. Above T_N, the Curie–Weiss behavior of (4.11) is fitted to the data. The extrapolations of the respective fits (*lines*) towards $1/\chi' = 0$ yield θ_{cw} which is shown in the case of $FeCl_2$ (*arrow*)

The huge susceptibility of $FeCl_2$ is a reminder of its 2d FM character at $T > T_N$. This becomes obvious on analyzing the Curie–Weiss type temperature dependence of χ' vs. T above T_N, which can be described according to [31, 32]

$$\chi = \frac{C}{T - \theta_{cw}}, \tag{4.11}$$

where C is the Curie constant and θ_{cw} is the Curie–Weiss temperature. In accordance with (4.11) the inset of Fig. 4.5 exhibits the virtual linear temperature dependence of $1/\chi'$ vs. T at $T > T_N$ for $FeCl_2$ (circles) and $Fe_{0.47}Zn_{0.53}F_2$ (squares), respectively. The corresponding Curie–Weiss temperatures $\theta_{cw} = +18.0$ and -31.7 K of $FeCl_2$ and $Fe_{0.47}Zn_{0.53}F_2$, respectively, are obtained from extrapolations of the linear regressions of $1/\chi'$ vs. T towards $1/\chi' = 0$, where formal divergence of the paramagnetic susceptibility sets in. Remarkably, the Curie–Weiss temperature of $FeCl_2$ is positive and, with a distance of 5.7 K, close to T_N. In accordance with the mean-field approximation where the parallel and the perpendicular susceptibility at T_N are given by (4.11), the distance between T_N and θ_{cw} is crucial for the absolute value of $\chi'(T_N)$. In addition, the Curie-constant C enters $\chi'(T_N)$ as a proportionality constant. It is determined from the inverse slope of the linear best fit of $1/\chi'$ vs. T (see inset of Fig. 4.5). Insertion of T_N, θ_{cw} and C into (4.11) yields $\chi'(T_N, FeCl_2) \approx 10.7\,\chi'(T_N, Fe_{0.47}Zn_{0.53}F_2)$, which is in qualitative agreement with $\chi'(T_N, FeCl_2)/\chi'(T_N, Fe_{0.47}Zn_{0.53}F_2) \approx 8$ determined from the susceptibility data at T_N (Fig. 4.5).

The microscopic interpretation of this phenomenological analysis is easily derived by inspection of the mean-field results

$$T_N = \frac{1}{3}S(S+1)\left(z_1\left|\frac{J_1}{k_B}\right| + z_2\left|\frac{J_2}{k_B}\right|\right) \tag{4.12}$$

and

$$\theta_{cw} = \frac{1}{3}S(S+1)\left(z_1\left|\frac{J_1}{k_B}\right| - z_2\left|\frac{J_2}{k_B}\right|\right), \tag{4.13}$$

where S is the spin quantum number, $J_{1,2}$ are the nearest and next-nearest-neighbor exchange constants and $z_{1,2}$ are the corresponding coordination numbers, respectively. (4.12) and (4.13) yield

$$\Delta T = T_N - \theta_{cw} = \frac{2}{3}S(S+1)z_2\left|\frac{J_2}{k_B}\right|. \tag{4.14}$$

In accordance with the 2d Ising character of $FeCl_2$ the AF exchange constant $|J_2|$ is small in comparison with $|J_1|$. Hence, ΔT is small in comparison with the corresponding value of $Fe_{0.47}Zn_{0.53}F_2$ where the AF exchange constant $|J_2|$ dominates the FM exchange $|J_1|$ [33]. In particular, $|J_2|$ of FeF_2 is about 10 times larger than the AF exchange constant of $FeCl_2$ [33]. As pointed out above, this yields $\chi'(FeCl_2) >> \chi'(Fe_{0.47}Zn_{0.53}F_2)$ in accordance with (4.11) and $C(FeCl_2) \approx C(Fe_{0.47}Zn_{0.53}F_2)$ (slopes of $1/\chi'$ vs. T inset of Fig. 4.5).

The subsequent Sects. 4.2.2 and 4.2.3 elaborate in detail the impact of the pinning layer susceptibility on the magnetic behavior of both types of heterosystems based on strong and soft antiferromagnets, respectively.

4.2.2 Exchange Bias in $FeF_2(001)/CoPt$

The perpendicular anisotropy of CoPt multilayers on the one hand and the Ising-type behavior of the strong antiferromagnet FeF_2 on the other hand make $FeF_2(001)/CoPt$ a prototypical heterostructure for the investigation of perpendicular exchange-bias [34, 35]. The uniaxial and exclusively perpendicular anisotropy reduces the degrees of freedom of the spin variables to a minimum. Hence, the complexity of possible spin structures is efficiently limited. In particular, complicated structures [36, 37, 38] which originate from spin-flop-like spin arrangements with canted AF interface moments and perpendicular orientation of the bulk *F*M moment relative to the AF easy axes are ruled out.

Moreover, heterostructures of FeF_2 single crystals covered with FM thin films are among the most frequently investigated and probably best-understood exchange-bias systems [43, 21, 40, 41, 42]. Large exchange-bias fields have been observed in the case of compensated AF (110) and (101) surfaces covered with Fe layers. Although the investigation of uncompensated surfaces like FeF_2 (001) seems to be straightforward, 90° coupling of the FM moments at the interface usually destroys the exchange-bias in accordance with (4.7). Hence, in order to avoid this problem and to take advantage of the above described simplification, here the uncompensated (001) surface of FeF_2 is covered with a FM CoPt layer of perpendicular anisotropy.

A sketch of the investigated $FeF_2(001)/(Co\,3.5\,\text{Å}/Pt\,12\,\text{Å})_3 Pt\,8\,\text{Å}$ multilayer is shown in Fig. 4.6. The rutile structure of the bulk antiferromag-

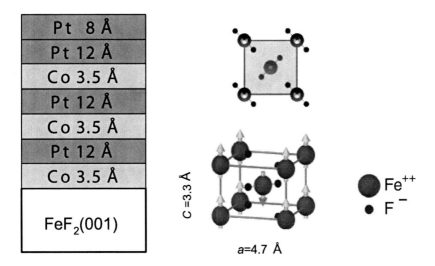

Fig. 4.6. Sketch of the $FeF_2(001)/(Co3.5\,\text{Å}/Pt12\,\text{Å})_3/Pt8\,\text{Å}$ heterostructure representing the prototypical system for perpendicular exchange-bias. The rutile structure of the bulk antiferromagnet consists of Fe^{2+} ions (*large spheres*) located on the body-centered tetragonal lattice with lattice parameters $a = b = 4.7$ and $c = 3.3\,\text{Å}$. Superexchange between Fe^{2+} ions of spin quantum number $S = 2$ is mediated by F^- ions (*small spheres*). *Arrows* indicate the AF spin ordering between ions at the center and the corners of the unit cell. A top view of the (001) surface exhibits details of the uncompensated spin structure

net , its corresponding spin structure and an additional top view of the (001) surface are shown in detail. Experimentally, the (001) orientation of the FeF_2 crystal has been checked by conoscopy using a polarizing microscope [44]. Before transferring the single crystal into the UHV chamber, the (001) plane is polished to optical flatness with 2.5μ m diamond paste. The $(Co\,3.5\,Å/Pt\,12\,Å)_3$ multilayer is deposited at 500 K under UHV conditions by thermal (Co) and electron-beam evaporation (Pt) onto the (001) surface of the FeF_2 crystal. The deposition rate is controlled by piezoelectric quartz resonators during the growth process. In addition, the thickness is determined by ex situ X-ray small-angle scattering. The result agrees with the nominal thickness within an error of 10%. Figure 4.7 shows exemplarily the result of the small-angle X-ray diffraction of a Pt 50 Å/(Co 3.5 ÅPt 12 Å)$_{15}$/Pt 8Å multilayer deposited on a glass substrate (compare Fig. 3.1).

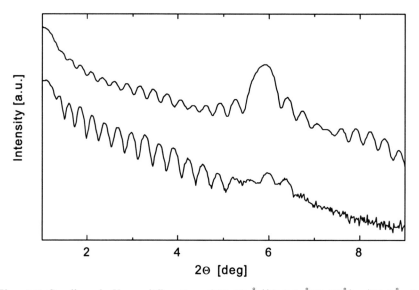

Fig. 4.7. Small-angle X-ray diffraction of Pt 50 Å/(Co3.5 ÅPt 12Å)$_{15}$/Pt 8Å deposited on a glass substrate [44]. The *lower* and *upper curves* display the measured and simulated intensities, respectively

The diffracted intensity is measured with a Philips (PW 1730) diffractometer at $\lambda = 1.542\,Å$ [44]. The lower curve of Fig. 4.7 displays the measured intensity as a function of 2θ, the angle between the incident and the reflected beams. The upper curve represents the result of a corresponding simulation involving the nominal layer thickness of Co and Pt, respectively [45]. The short period of the intensity oscillations originates from interference involving the total thickness of the metallic layer, while the superlattice peak at $2\theta \approx 6°$ originates from the artificial periodicity which is given by the thickness of the stacked Co/Pt bilayers.

In order to prevent oxidation of the heterostructure, the last Pt layer is covered by an additional Pt layer of 8 Å thickness. By this measure, ex-situ investigation of the heterosystem becomes possible. Despite their rather complex structure, the usefulness of Co/Pt multilayers becomes obvious when taking into account the advantages of combining perpendicular anisotropy on the one hand with a high magnetic moment on the other hand [46]. The latter causes an adequate contrast between the magnetization of the FM layer and the large field-induced magnetization of the bulk antiferromagnet. For example, in Fe layers perpendicular magnetic anisotropy is possible only in the ultra-thin limit [47, 48].

In order to investigate the perpendicular exchange-bias effect, magne-tometric measurements were done with a commercial 5 T SQUID system (Quantum Design MPMS5S). Hysteresis loops are performed after heating the sample to 200 K and subsequent cooling to 10 K in various applied axial magnetic freezing fields. Figure 4.8 shows a typical hysteresis loop (circles) at 10 K after cooling the heterosystem to below the Néel temperature of $T_N = 78.4$ K in an axial magnetic field of $\mu_0 H = 0.2$T. The magnetic phase diagram of FeF_2 is depicted in the inset.

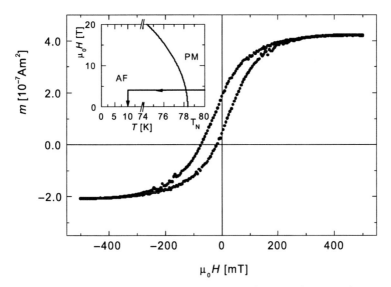

Fig. 4.8. Hysteresis loop of the $FeF_2(001)/(Co\ 3.5\ Å/Pt\ 12\ Å)_3/Pt\ 8Å$ system at $T = 10$ K for the freezing field of $\mu_0 H = 0.2$ T. *Solid curve* in the *inset* represents the second order phase transition line separating the PM from the AF phase. *Arrows* indicate a typical field-cooling procedure from the PM into the AF phase to below $T_N = 78.4$ K

The solid curve represents the second order phase transition line which separates the PM from the AF ordered phase. The critical line $T_c(H) = 78.4$

K-9.95×10^{-3} K/T^2 $(\mu_0 H)^2$ has been determined from ultrasonic attenuation measurements in high steady magnetic fields $0 < \mu_0 H < 20$ T produced by a Bitter-type solenoid [49]. The arrows indicate a typical field-cooling procedure from the PM into the AF phase which prepares the spin structure of the heterosystem.

While the exchange-bias shifts the hysteresis loop along the field axis, a shift along the axis of the magnetic moment is encountered in FeF$_2$/FM systems in general [40]. It is due to piezomagnetism, which is allowed by symmetry in rutile-type AF compounds [50] and may be induced by residual shear-stress [51]. In accordance with (4.1) exchange-bias requires a net AF interface moment. While its origin seems to be obvious in the case of uncompensated AF surfaces, piezomagnetism provides the possibility to generate an AF interface moment in the case of compensated AF surfaces. Details of the impact of piezomagnetism on the exchange-bias are discussed in Sect. 4.3, where heterostructures with compensated AF interfaces and planar anisotropy are investigated.

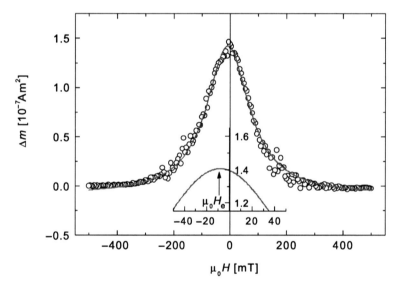

Fig. 4.9. Difference curve Δm vs. $\mu_0 H$ between ascending and descending branches of the hysteresis loop shown in Fig. 4.8. The *solid line* represents the corresponding best fit to the Lorentzian function of (4.18). It is shifted by the exchange-bias field $\mu_0 H_e = -8.5$ mT. The *inset* exhibits the vicinity of the maximum in detail. The shift of the maximum position is indicated by an *arrow*

In order to safely determine the exchange-bias under the influence of the piezomagnetic shift, the hysteresis loops are analyzed by subtracting the descending and ascending branches, thus eliminating both the constant piezomagnetic and the large moment of the FeF$_2$ crystal which is proportional to the magnetic field. Note that the data in Fig. 4.8 are corrected already with respect to the magnetic moment induced by the linear AF susceptibility of FeF_2. Figure 4.9 shows the field dependence of the difference moment $\Delta m = m_d - m_a$ obtained after subtracting the descending and ascending branches, $m_{d,a}$. Only hysteretic effects contribute to this difference data Δm vs. $\mu_0 H$. In the case of $FeF_2/CoPt$ heterostructures they are exclusively attributed to the FM CoPt layer. In particular, Δm vs. $\mu_0 H$ is shifted along the magnetic field axis in accordance with the exchange-bias effect. It is straightforward to show that this shift is precisely given by the exchange-bias field $H = H_e$. The branches $m_{d,a}$ are separated into the reversible background term, $b(H)$, and the corresponding remainders which give rise to the FM hysteresis. Formal expansion of the latter terms into power series in the vicinity of the coercive fields H_{c1} and H_{c2}, respectively, yields

$$m_{d,a} = \sum_{n=0}^{\infty} a_n (H - H_{c1,c2})^{2n+1} + b(H), \qquad (4.15)$$

where point symmetry at H_{c1} and H_{c2} for $m_{d,a} - b(H)$ and symmetric magnetization-reversal between the descending and the ascending branches are assumed. The latter condition gives rise to identical expansion coefficients a_n for m_d, and m_a, respectively. In accordance with (4.15) the derivative of $\Delta m = m_d - m_a$ with respect to H reads

$$\frac{\partial \Delta m}{\partial H} = \sum_{n=0}^{\infty} (2n + 1)a_n [(H - H_{c1})^{2n} - (H - H_{c2})^{2n}]. \qquad (4.16)$$

Inspection of (4.16) shows that the condition

$$(H - H_{c1})^2 = (H - H_{c2})^2 \qquad (4.17)$$

yields $\partial \Delta m/\partial H = 0$. Hence, the solution of (4.17) provides the maximum position of $\Delta m(H)$ at $H_{max} = (H_{c1} + H_{c2})/2$, which is just the conventionally determined exchange-bias field in accordance with (4.5).

A best fit of the Lorentzian function

$$L(H) = \frac{a}{4(H - H_e)^2 + w^2} \qquad (4.18)$$

to the Δm vs. $\mu_0 H$ data yields the amplitude a, the width w and, in particular, the shift H_e of Δm vs. $\mu_0 H$ with respect to $H = 0$.

The subsequent analysis demonstrates that the resulting exchange-bias field, H_e, does not depend on the choice of the fitting function (4.18) which, although it fits the data pretty well, has no and needs no further physical justification. In order to check the unambiguous determination of H_e from a best fit of (4.18) to the Δm vs. $\mu_0 H$ data one has to prove that

$$\frac{\partial}{\partial \tilde{H}} \int\limits_{H_e-\Delta H}^{H_e+\Delta H} \left(\Delta m(H) - \frac{a}{4(H-\tilde{H})^2 + w^2} \right)^2 dH \bigg|_{\tilde{H}=H_e} = 0 \qquad (4.19)$$

in the case $\Delta m(H - H_e) = \Delta m(-H + H_e)$. Here $\Delta m(H)$ is the unknown ideal theoretical function that provides the best possible description of the data. Substitution of $h = H - H_e$ into (4.19) and subsequent calculation of the derivative with respect to $\tilde{H}$ yields

$$2 \int\limits_{-\Delta H}^{\Delta H} \left(\Delta m(h) - \frac{a}{4h^2 + w^2} \right) \frac{8ah}{(4h^2 + w^2)^2} dh = 0. \qquad (4.20)$$

Inspection shows that (4.20) is identically fulfilled in accordance with the general criterion that the integration of an odd function within the symmetric interval yields zero. In particular, the result is independent of the values of the parameters a and w and the details of the even function $\Delta m(h)$. Hence, a best fit of (4.18) unambiguously provides H_e. It is obvious that also other ansatz functions like, e.g., the Gaussian fitting function yield the same result. In contrast with the conventional determination of the exchange-bias field by calculating $H_e = (H_{c1} + H_{c2})/2$ from the intercepts $H_{c1,c2}$ of $M(H)$ with the H axis, the above method to extract H_e involves the data of the entire hysteresis loop and is, hence, assumed to be more accurate. Moreover, measurements which do not determine the absolute value of the magnetic moment may give rise to meaningless exchange-bias fields when analyzing the hysteresis loops in the conventional way. Any loop shift along the moment axis creates an apparent exchange-bias in the case of loops which deviate from rectangular shape with infinite slope of m vs. H at $H_{c1,c2}$.

A systematic study of the freezing field dependence H_e vs. H_F is shown in Fig. 4.10. The exchange-bias field is largest for small freezing fields $\mu_0 H_e(0.2\,T) = -8.5$ mT, whereas it decreases considerably in a high-field $\mu_0 H_e(5\,T) = -2.2$ mT. In the weak H_F limit $|H_e|$ shows a steep increase with increasing freezing field up to a maximum at $H_F \approx 0.1$T. Then it decreases with increasing H_F and nearly vanishes for fields above 2T. It is noticed that the value of H_e remains negative, in contrast with recent results on FeF_2 -based systems with planar anisotropy [21, 52].

It should be stressed that the measurements were done in arbitrary order at different values of H_F in order to avoid any errors due to training effects, although no effect of the number of measurements on the value of the exchange-bias could be observed. This was proven by a series of measurements with constant freezing field. Within the errors of the exchange-bias field no significant effect was observed.

Similar drastic changes of H_e vs. H_F as shown in Fig. 4.10 were reported previously for FeF_2–Fe bilayers exhibiting in-plane anisotropy [52]. Qualitatively the occurrence of positive exchange-bias was explained by a competition between the AF-FM exchange interaction and the coupling energy

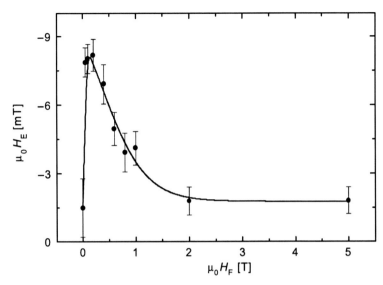

Fig. 4.10. Dependence of the exchange-bias on the freezing field in a $FeF_2(001)/(Co$ 3.5 Å/Pt 12 Å)$_3$/Pt 8Å multilayer. Experimental data (*circles with error bars*) and their best fit to (4.29) (*solid line*) are shown

between the AF surface layer and the magnetic field, H_F. In weak freezing fields the AF-FM exchange prevails over the Zeeman energy gained by the AF interface layer in the applied field. The dominant AF exchange coupling results in the usual negative exchange-bias, $H_e < 0$. However, in high freezing fields and under the constraint of AF interface exchange interaction it may happen that the Zeeman energy overcomes the exchange energy. In that case, the FM and the AF topmost layers become aligned with H_F. When freezing-in this unfavorable spin configuration, this will give rise to backswitching of the FM magnetization in zero external field, hence producing a positive exchange-bias, $H_e > 0$.

In order to describe quantitatively the strong dependence of the exchange-bias on the freezing field H_F as shown in Fig. 4.10 or previously [52] one has to consider the actual spin structure at the AF/FM interface encountered at the Néel temperature, below which the AF domain structure is established. From the above consideration one may anticipate that intermediate H_e values will occur in intermediate freezing fields H_F, which do not completely align either the FM or the AF spins at the interface.

Since it is known that the exchange-bias effect depends on the spin structure at the AF/FM interface, or more exactly, on the energy gained by exchange interaction, $\sum_{i,j} J_{ij}\sigma_i^{AF}\sigma_j^{FM}$ [9, 10, 24], it will be useful to consider equilibrium thermodynamics of the interface at $T \approx T_N << T_c (= FM$ Curie temperature) under the constraint of a fixed external field, H.

The Curie temperature $T_c \approx 600$ K [53, 54] of the FM Co/Pt-multilayer is much higher than $T_N = 78.4$ K of the antiferromagnet FeF$_2$. Hence, the Co/Pt- multilayer is magnetically ordered at all temperatures $T < T_c$. In the case of a uniaxial ferromagnet, at $T << T_c$, the net magnetic moment is determined by its domain structure and can be expressed by the number of up and down magnetic moments. Hence, the total magnetic moment of the FM interface layer reads

$$S_{FM} = (n_{FM}^+ - n_{FM}^-)m_{FM} = (2n_{FM}^+ - n_0)m_{FM}, \qquad (4.21)$$

where m_{FM} is the magnetic moment per atom and $n_0 = n_{FM}^+ + n_{FM}^-$ is the total number of FM interfacial up and down moments, $n_{FM}^{\pm}$. In order to model the field dependence of the magnetization of the ferromagnet in the absence of exchange-bias, i.e. at $T_N < T < T_c$, we make the simple linear ansatz

$$n_{FM}^+(H) = \begin{cases} 0 & for \quad H < -H_s, \\ (n_0/2)(1 + H/H_s) & for \;\; -H_s < H < H_s, \\ n_0 & for \quad H > H_s, \end{cases} \qquad (4.22)$$

where H_s is the saturation field value. Note that the exchange-bias of the magnetization-reversal curve does not explicitly depend on the width of the hysteresis loop. Therefore, for sake of simplicity hysteresis is completely neglected and homogeneous domain nucleation and growth behavior is assumed. A typical situation encountered for $0 < H < H_s$ is depicted in Fig. 4.11 (above the interface).

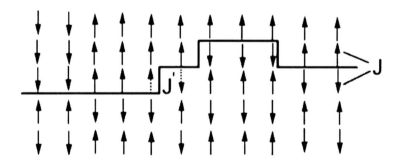

Fig. 4.11. Schematic spin structure at the FM/AF interface (*line*) and its adjacent layers after cooling to below T_N in an axial freezing field $0 < H < H_s$. *Arrows* indicate the magnetic moments of the FM (*above the interface*) and AF system (*below the interface*), respectively. At the interface adjacent spins are coupled by the AF exchange constant J along the spin direction (*solid arrows*) and, in addition, by J' (*dashed arrows*) at the steps of the interface

Since the magnetization-reversal is strongly supported by the perpendicular anisotropy of the Co/Pt multilayer system [53, 54], saturation is achieved at fairly low values of H_s. These are definitely lower than $H_s = \mu_B$ (in SI units), which would be expected for a FM thin-film with mere shape anisotropy when exposed to a perpendicular magnetic field. Owing to the curvature of the experimental M vs. H curves (see e.g. Fig. 4.8) it seems reasonable to treat H_s as a fitting parameter. Moreover, it is assumed that the spin structure of the ferromagnet which establishes on applying $H = H_F$ at $T > T_N$ is not affected by the ordering process of the exchange-coupled antiferromagnet on cooling to below T_N, although the unidirectional anisotropy originates from this coupling. Hence, the number of up spins n_{FM}^+ as given by (4.22) does not change on cooling towards T_N and below. Under this constraint it is now possible to calculate the spin structure and magnetization of the interface layer of the antiferromagnet within the framework of equilibrium thermodynamics . To this end the partition function is written down by taking into account the four possible configurations of the AF/FM spin pairs at the interface. They originate from the combinations of the spin values $\sigma_{FM} = \pm 1$ and $\sigma_{AF} = \pm 1$, respectively. The energy function takes into account the exchange and the Zeeman energies . The latter affects only σ_{AF}, because the orientation of σ_{FM} is assumed to be independent of temperature at $T \ll T_c$ for a given freezing field. Hence, no thermodynamic consideration of σ_{FM} is necessary. Moreover, it is assumed that on cooling the ordering of the spins in the AF system starts at the AF/FM interface at $T = T_N + \delta T$, where $0 < \delta T \ll T_N$.

This behavior is reasonable, because similar proximity effects have been observed on EuS precipitated in Co and, in particular, in FeF_2–Fe bilayers [55, 56]. Therefore, exchange coupling between σ_{FM} and σ_{AF} at the interface is taken into account, but neglected between σ_{AF} and its AF bulk neighbors. Their ordering requires further cooling to $T \leq T_N$.

Once the spin structure at the interface is stabilized, the underlying antiferromagnet develops its domain structure from the interface into the bulk of the crystal. This final domain structure of the AF gives rise to the unidirectional anisotropy, which characterizes the exchange-bias. Within these approximations the energy function, which controls the spin arrangement of the interface, reads

$$E = -J\sigma_{FM}\sigma_{AF} - g\mu_B\mu_0 H\sigma_{AF}. \tag{4.23}$$

For a given orientation of σ_{FM} there are two different states $\sigma_{AF} = \pm 1$. This yields two partition functions $Z^\pm$ for $\sigma_{FM} = \pm 1$, respectively. They read

$$Z^\pm = 2\cosh\left((\pm J + g\mu_B\mu_0 H)/k_B T\right) \tag{4.24}$$

and allow us to calculate the thermally averaged magnetizations

$$m^\pm = \frac{1}{\mu_0}\frac{\partial}{\partial H}k_B T \ln Z^\pm = g\mu_B \tanh\left((\pm J + g\mu_B\mu_0 H)/k_B T\right). \quad (4.25)$$

Up to now no interface roughness has been taken into account. Let us now consider steps at the interface, which give rise to a new kind of interlayer exchange coupling J' between σ_{FM} and σ_{AF} (Fig. 4.10, broken arrows). Since the steps may be regarded as (100) or (110) planes, which give rise to strongly enhanced exchange-bias (e.g. $\mu_0 H_e \approx 50$ mT for FeF_2 (100)/Fe [102]) we anticipate $|J'| >> |J|$. Hence, those spins which are coupled via J' to σ_{AF} do not participate in the above thermodynamic consideration, but are rigidly coupled to their neighboring FM spin.

Let $n_s^\pm$ be the number of spins σ_{FM} located at steps and $\tilde{n}_{FM}^\pm$ the remaining spins on the terraces. The magnetic moment S_{AF} of the AF interface layer (Fig. 4.11, solid line) then reads

$$S_{AF} = \tilde{n}_{FM}^+ m^+ + \frac{J'}{|J'|}n_s^+ m_0 + n_{FM}^- m^- - \frac{J'}{|J'|}n_s^- m_0, \quad (4.26)$$

where m_0 is the absolute value of the magnetic moment of the spin in the AF system. With $n_{FM}^\pm = \tilde{n}_{FM}^\pm + n_s^\pm$, $n_0 = n_{FM}^+ + n_{FM}^-$ and the topographic roughness parameter $\alpha = n_s^\pm/n_{FM}^\pm$, one obtains

$$S_{AF} = n_{FM}^+(1-\alpha)\left[m^+ - m^-\right] + n_0(1-\alpha)m^- + \frac{J'}{|J'|}\alpha m_0(2n_{FM}^+ - n_0) \quad (4.27)$$

with $0 \le \alpha \le 1$. This magnetization of the AF topmost layer is assumed to determine the domain structure of the bulk AF substrate (Fig. 4.11, below the interface) and to remain invariant both on cooling to $T << T_N$ and upon cycling between positive and negative saturation of the FM subsystem.

From the generalized Meiklejohn–Bean considerations of Sect. 4.1 the exchange-bias field H_e is given by

$$H_e \propto [(1-\alpha)J + \alpha J']S_{AF}S_{FM} \quad (4.28)$$

in the case of Ising anisotropy of the antiferromagnet . Although the Stoner-Wohlfarth model of coherent rotation is at a first glance not appropriate in the case of Ising anisotropy of the ferromagnet, the magnetization-reversal via domain formation can be mapped onto a rotation process where the z-component of the FM moment changes from its positive to the negative saturation value. In order to take into account the thermal control of the evolution of the interface magnetic moments during the freezing process, the $T = 0$ saturation value S_{AF} of (4.1) is replaced by the thermal equilibrium values of (4.27). With $\Delta m \equiv m^+ - m^-$ and $S_{FM} = (2n_{FM}^+ - n_0)m_{FM}$ one finally obtains

$$H_e \propto J\left[(1-\alpha) + \alpha J'/J\right](2n_{FM}^+ - n_0)m_{FM}\left(n_{FM}^+(1-\alpha)\Delta m\right.$$
$$\left. + n_0(1-\alpha)m^- + J'/|J'|\alpha m_0(2n_{FM}^+ - n_0)\right). \tag{4.29}$$

Expression (4.29) is now best-fitted to the H_e vs. H_F data by inserting (4.22) with $H = H_F$ and (4.25) by letting $T = T_N = 78$ K. Here we assume that the spin structure at the interface, which establishes at $T = T_N$, will not change when cooling to the measurement temperature, $T = 10$ K. Hence, all interactions both within the AF interface layer and between this layer and the ordered bulk are neglected. $J/k_B T_N$, H_s, α, $J'/|J'|$ and $g\mu_B/k_B T_N$ enter (4.29) as fitting parameters, while n_0, m_{FM} and $J[(1-\alpha)+\alpha J'/J]$ become a part of the proportionality constant which transforms (4.29) into an equation. A constant offset is added during the fitting procedure in order to describe the background of the H_e vs. H_F curve as defined by the data point at $H_F = 0$. It originates from the remanent magnetization of the FM layer which gives rise to a preferential antiparallel alignment of S_{AF} on freezing in accordance with the AF exchange coupling at the interface. The result of the best fit is shown in Fig. 4.10 (solid line). The data are well described within their error bars. As expected, the microscopic exchange parameter $J/k_B T_N = -0.46$ is negative. Hence, AF coupling at the interface is favored. The maximum of the H_e vs. H_F curve and, in particular, the decrease of H_e with increasing freezing field originates from the competition between this AF exchange and the Zeeman energy. Moreover, $\mu_0 H_s = 0.12$T is in acceptable agreement with the saturation field revealed by the hysteresis loop of Fig. 4.8, where about 80% of μ_B is achieved at $\mu_0 H_s = 0.12$T. The steep increase of H_e vs. H_F in the weak-field limit, $H_F < H_s$, corroborates the rapid saturation of the FM overlayer. The subsequent decrease of H_e is understood by the weakening of the AF ordering owing to dominance of the Zeeman energy after reaching full FM saturation. The best-fitted value $\alpha = 0.2$ indicates a rough interface, where 20% of the FM spins are located at step positions. This roughness very probably originates from the mechanical polishing procedure of the (001) surface of FeF_2. Note that the proportionality (4.29) opens the possibility of positive exchange-bias by reduction of α towards zero, which has the physical meaning of a perfect interface. Then expression (4.29) yields

$$H_e \propto J(2n_{FM}^+ - n_0)m_{FM}(n_{FM}^+\Delta m + n_0 m^-), \tag{4.30}$$

which gives a continuous crossover from negative to positive exchange-bias upon increasing the freezing field H_F. The proportionality constant which contains the product of $J[(1-\alpha)+\alpha J'/J]$, n_0 and the magnetic moment of the Co atoms, m_{FM}, reads $P = 0.11$ mT. Further, the coupling constant J' also turns out to be negative, since $J'/|J'| = -1$ is required to obtain the appropriate shape of the fitting function. The best-fitted value $g\mu_B/k_B T_N = 1.24 T^{-1}$ is by a factor of 65 higher than expected from the single magnetic moment, $g\mu_B$, of Fe^{2+} in bulk FeF_2 on letting $g = g_\| = 2.2$ [57]. This discrepancy is probably due to the neglect of any correlation between

the AF interface and bulk spins. In a first approximation the above number corresponds to a cluster size of about four nearest-neighbor distances,
which appears reasonable for an antiferromagnet just above T_N. These clusters effectively enhance the field contribution which enters the weight factor
$\tanh((\pm J + g\mu_B\mu_0 H)/k_B T_N)$ in (4.25). They thus determine the magnetic
structure of the interface, which will not change upon cooling. Measurements
at $T = 10$ K are, hence, able to provide information on the spin ordering
taking place at T_N.

In addition to $FeF_2(001)/(Co\,3.5\,\text{Å}/Pt\,12\,\text{Å})_3/Pt\,8\text{Å}$ the freezing field dependence of the exchange-bias of $FeF_2(001)/Pt\,15\,\text{Å}(Co\,3.5\,\text{Å}/Pt\,12\,\text{Å})_5/Pt\,8\,\text{Å}$
has been investigated [58]. This heterostructure contains an additional Pt
buffer of 15Å thickness which separates the antiferromagnet from the subsequent $(Co\,3.5\,\text{Å}/Pt\,12\,\text{Å})$ multilayer. Qualitatively, the influence of the
Pt buffer on the freezing field dependence is marginal. The H_e vs. H_F -
data (open circles) and the corresponding fit of (4.29) (line) are shown
in the inset of Fig. 4.12. Quantitatively, the absolute value of the exchange parameter $J/k_B T_N = -0.11$ is reduced by 81% with respect to
$FeF_2(001)/(Co\,3.5\,\text{Å}/Pt\,12\,\text{Å})_3/Pt\,8\,\text{Å}$ in accordance with the expected decay of the exchange interaction which takes place with increasing distance
between Co moments and the AF interface. The remaining parameters of the
best fit read $\mu_0 H_s = 0.08$ T, $\alpha = 0.64$ and $g\mu_B/k_B T_N = 1.15$ T^{-1}. The values of the saturation field and the magnetic moment are in good agreement
with the previous findings. The strong increase of the roughness parameter is
reasonable when taking into account a crystal polishing procedure of minor
quality.

Moreover, the temperature dependence of the exchange-bias at constant
freezing field $\mu_0 H_F = 0.1$ T is investigated by SQUID magnetometry. Figure
4.12 shows the H_e vs. T data obtained from the corresponding hysteresis loops
at temperatures $2K \leq T \leq 50$ K after cooling the system in the axial freezing
field $\mu_0 H_F = 0.1$ T, which maximizes the exchange-bias effect (inset Fig.
4.12). The temperature dependence of H_e can be modeled within a mean-
field approach generalizing expression (4.29).

For that purpose, the field dependence of $m^{\pm} = g\mu_B \tanh((\pm J + g\mu_B\mu_0 H)$
$/k_B T)$ is extended by a molecular field

$$H_m \propto |T - T_N|^{\beta}, \tag{4.31}$$

where β has been fixed during the fitting process as $\beta = 0.325$, which is the
3d Ising critical exponent of the order-parameter. In a first approximation it
takes into account the interaction of the AF interface with the subsequent
layers of the AF bulk. The molecular field stabilizes the spin structure of the
AF interface and, hence, supports the exchange-bias field. With increasing
temperature H_m decreases which gives rise to a decrease of H_e. Although
it is obvious that this approach is oversimplified, it gives, some insight into
the basic mechanism of the temperature dependence of the exchange-bias.

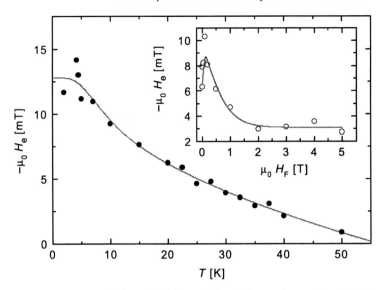

Fig. 4.12. Experimental data (*solid circles*) of the exchange-bias H_e for various temperatures T in $FeF_2(001)/Pt\,15\,\text{Å}(Co\,3.5\,\text{Å}/Pt\,12\text{Å})_5/Pt\,8\text{Å}$ and the corresponding best fit of (4.29) extended by the molecular field term H_m (*solid line*). *Inset* shows the freezing field dependence of the exchange-bias (*open circles*) obtained from hysteresis loops m vs. $\mu_0 H$ at $T = 10$ K for various freezing fields H_F and the corresponding best fit of (4.29) (*solid line*)

It originates from the temperature dependence of the AF interface magnetic moment S_{AF} in accordance with the simple MB formula (4.5). S_{AF} establishes on cooling to below T_N in a given freezing field under the influence of the exchange field of the FM top layer and the temperature dependent molecular field of the bulk antiferromagnet.

4.2.3 Domain-state Susceptibility in FeCl$_2$/CoPt-Heterostructures

The previous section exhibited the magnetic behavior of the perpendicular exchange-bias system FeF$_2$–CoPt. As pointed out before, this type of heterostructure is characterized by its underlying strong antiferromagnet, where a large AF superexchange interaction gives rise to a stable freezing of the AF spin structure on cooling to below T_N. At low temperatures, $T << T_N$, it pins the ferromagnet while the AF spin arrangement remains virtually unaffected from the magnetization-reversal process of the FM layer during the hysteresis loop.

However, this ideal situation changes when the energy barriers involved in the stabilization of a spin configuration decrease or the spin rearrangement is thermally assisted on approaching T_N. In these cases, a magnetization-reversal process of the exchange-coupled ferromagnet gives rise to a partly irreversible evolution of the AF spin structure [59, 60]. As pointed out in Sect. 4.2.1, $FeCl_2$ is a soft antiferromagnet according to its weak AF inter layer interaction. Moderate axial magnetic fields of $\mu_0 H \approx 1$ T give rise to a metamagnetic transition (compare Fig. 2.5 with inset Fig. 4.8).

Hence, $FeCl_2(111)/$ (Co 3.5 Å Pt 12Å)$_{10}$ /Pt 8Å reflects a prototypical system where excessive AF domain growth and evolution can be studied. In the following, it will be shown that in heterostructures based on the soft antiferromagnet $FeCl_2$ the long-range-ordered AF state breaks down into a metastable domain state on field-cooling. This domain state is highly susceptible to axial magnetic fields. Moreover, upon heating towards T_N it exhibits a thermally activated relaxation towards the AF long-range-ordered ground state. Figure 4.13a and b show the temperature dependence of the ac-susceptibility $\chi = \chi' - i\chi''$ of a $FeCl_2(111)/$ (Co 3.5 Å Pt 12Å)$_{10}$ /Pt 8Å multilayer as measured by SQUID susceptometry at the frequency $f = 10$ Hz [61]. The preparation of the initial state of the sample is schematically shown in the magnetic phase diagram of $FeCl_2$ as depicted in the inset of Fig. 4.14. H designates the magnetic field applied perpendicularly to the (111) cleavage plane. It gives rise to isothermal metamagnetic AF-to-PM phase transitions below T_N, involving a mixed phase (M) along the low-T first-order phase line.

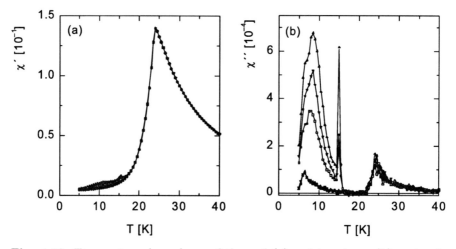

Fig. 4.13. Temperature dependence of the real (**a**) and imaginary (**b**) parts of the ac susceptibility $\chi = \chi' - i\chi''$ of $FeCl_2(111)/(Co\ 3.5\ Å/Pt\ 12\ Å)_{10}/Pt\ 8\ Å$ at frequency $f = 10$ Hz. Details of the excess susceptibility and the preparation of the initial state are shown and described in Fig. 4.14

After cooling in an axial freezing field of $\mu_0 H = 5$ T to $T_0 = 11$ (up triangles), 12 (down triangles) and 13 K (squares), respectively, the field is rapidly decreased towards zero where the AF order of the pinning layer becomes frozen-in. On subsequent heating, both χ' and χ'' vs. T exhibit pro-

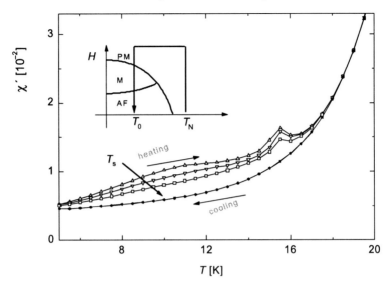

Fig. 4.14. Schematic phase diagram of $FeCl_2$ (*inset*) shows the preparation of the initial state. After freezing in an axial field of $\mu_0 H = 5$ T down to $T_0 = 11$ (*up triangles*), 12 (*down triangles*) and 13 K (*squares*), respectively, the field is rapidly decreased towards zero and χ' vs. T (*open symbols*) is measured for $T_s = 5$ K$< T < T = 40$ K to above $T_N = 23.7$ K. *Solid symbols* show χ' vs. T on subsequent cooling

nounced excess contributions showing a broad and a narrow peak at $T \approx 12$ K and ≈ 15 K, respectively. They vanish on subsequent zero-field cooling from $T = 40$ K back to $T = 5$ K, i.e. to far below the Néel temperature $T_N = 23.7$ K of $FeCl_2$ (see details of χ' vs. T in Fig. 4.14).

Figure 4.15 shows a sketch of the spin structure of a $FeCl_2/CoPt$ heterolayer growing at a single-atomic step of the as-cleft $FeCl_2$ single crystal. Such a structure is expected on cooling to below the Néel temperature in an axial magnetic field that aligns the FM moments, but is small in comparison with the exchange field between the AF and FM spins at the interface. In that case, the interface coupling (assumed to be AF in Fig. 4.15) controls the preferred orientation of the AF moments at the interface. Its roughness prevents the evolution of long-range AF order and gives rise to the formation of AF domains. They start to grow at the interface and end up in the AF bulk.

Note that a similar domain state is expected when $FeCl_2$ without a FM top layer is isothermally demagnetized from the saturated PM into the AF state across the spin-flip transition (inset Fig. 4.15). In that case, the axial magnetic field at the spin-flip transition aligns the moments of the uncompensated (111) surface parallel to the c axis. These spins are particularly affected by the field owing to the reduced superexchange interaction which surface moments experience with respect to the bulk moments of the anti-ferromagnet . Hence, in the case of $FeCl_2$ single crystals the applied field corresponds to the exchange field at the AF/FM interface of the $FeCl_2$/CoPt heterostructure. The latter, however, is supposed to exceed the applied field when the spin-flip transition takes place in a heterostructure in the presence of exchange at the interface [18].

Analogously to the situation shown in Fig. 4.15 the field-aligned upper-most layer at the rough surface is the starting point for AF domain formation [62, 63]. Figure 4.15 shows two types of domain walls. On the one hand, the 2d FM order within the (111) planes is broken. The basic elements of these 'vertical' domain walls are pairs of antiparallel aligned spins (see the gray-highlighted spin pair in Fig. 4.15). They enhance the magnetic energy by the individual amount of $+2J$ each, where $J > 0$ is the intraplanar FM exchange constant.

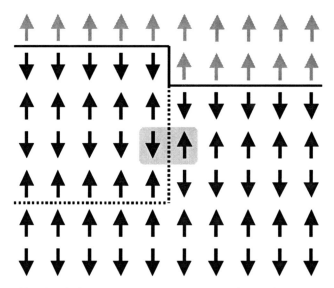

Fig. 4.15. Sketch of the spin structure of a $FeCl_2$/CoPt heterostructure after freezing in an axial magnetic field that polarizes the FM moments (uppermost layers). The *solid line* represents the AF/FM interface. Interface roughness drives the AF bulk via exchange interaction into the domain state. Pairs of spins (*gray background*) build up the vertical domain-wall (*vertical dashed line*). AF stacking faults give rise to a horizontal domain-wall perpendicular to the c axis (*horizontal dashed line*)

On the other hand, AF stacking faults give rise to 'horizontal' AF domain walls which break the symmetry along the c axis and, hence, carry a net magnetic moment when forming under an aligning external field [64]. Both types of domain walls correspond to distinct susceptibility contributions and will be discussed subsequently.

As the starting point the contribution of the spin pairs within the vertical walls is investigated. Each pair gives rise to a susceptibility contribution which is determined by the energies of the four spin configurations of the two Ising spins. Assuming that the AF ordered neighborhood of a given spin pair remains virtually unchanged in the temperature range $5\,\text{K} < T < 20$ K where the excess susceptibility evolves, the four energy values

$$E_j = K_j - g\mu_B\mu_0 H S_j^{\text{tot}} \tag{4.32}$$

describe the equilibrium thermodynamic behavior of the spins. K_j summarizes the exchange interaction energy between the spins of a given pair and all the surrounding spins that interact with the pair via exchange.

The second term takes into account the Zeeman energy of the two spins in the applied magnetic field. The index j labels one out of the four configurations which correspond to the total spin values $S_j^{\text{tot}} = \pm 2$ in the case of parallel spin alignment ($j = 0$ and 1), while the two spin configurations of antiparallel alignment yield $S_j^{\text{tot}} = 0$ ($j = 2$ and 3). Each spin pair is assumed to give rise to the same susceptibility contribution. The total susceptibility is then calculated from the total free-energy expression of the spin pairs,

$$F_{\text{sp}} = -k_B T N \ln\left(\sum_{j=0}^{3} e^{-E_j/(k_B T)}\right), \tag{4.33}$$

where N is the number of spin pairs building up the AF domain walls. The static magnetic susceptibility in zero external field, $H = 0$, is calculated according to

$$
\begin{aligned}
\chi_{\text{sp}} &= -\left(\frac{\partial^2 F_{\text{sp}}}{\partial H^2}\right)_{H=0} \\
&= C\beta \frac{e^{\beta(\tilde{K}_1+\tilde{K}_2+\tilde{K}_3)}(4e^{\beta\tilde{K}_1} + e^{\beta\tilde{K}_2} + e^{\beta(\tilde{K}_1+\tilde{K}_2)} + e^{\beta\tilde{K}_3} + e^{\beta(\tilde{K}_1+\tilde{K}_3)})}{(e^{\beta(\tilde{K}_1+\tilde{K}_2)} + e^{\beta(\tilde{K}_1+\tilde{K}_3)} + e^{\beta(\tilde{K}_2+\tilde{K}_3)} + e^{\beta(\tilde{K}_1+\tilde{K}_2+\tilde{K}_3)})^2},
\end{aligned}
\tag{4.34}
$$

where $\tilde{K}_j = K_j - K_0$, $\beta = 1/(k_B T)$ and $C = N(2g\mu_B\mu_0)^2$. On a mesoscopic scale, each domain-wall separates two regions which are related by time inversion (see Fig. 4.15). Hence, the two configurations with $S_{j=0,1}^{\text{tot}} = \pm 2$ possess identical exchange energies, i.e. $K_0 = K_1$. Let $j = 3$ label the spin configuration with $S_{j=3}^{\text{tot}} = 0$ where both spins of the pair are flipped with respect of the configuration $j = 2$ which is shown in Fig. 4.15. In that case, $j = 3$ represents the energetically most unfavorable state. Hence, it is reasonable to assume that its thermal excitation is negligible in comparison with the population of the states $j = 0, 1, 2$. Therefore, the susceptibility is determined by the single energy parameter $\tilde{K}_2$, the difference between the exchange energy of the $S_{j=0,1}^{\text{tot}} = \pm 2$ and $S_{j=2}^{\text{tot}} = 0$ configurations. The simplified expression (4.34) then reads

$$\chi_{\text{sp}} = \frac{2C\beta}{2 + e^{\beta \tilde{K}_2}}. \tag{4.35}$$

The domain structure is a metastable state which relaxes into the AF long-range-ordered state. Its relaxation time is long in comparison with the inverse ac frequency of the measurement $\tau = 1/\nu = 0.1$ s. Therefore, the relaxation affects the susceptibility only by the reduction of the number of spin pairs which enter (4.35) via C. The expected decay rate, dN/dt, of the spin pairs is given by

$$\frac{dN}{dt} = -\alpha(T)N, \tag{4.36}$$

where a decay constant with Arrhenius-type thermal activation,

$$\alpha = \alpha_0 e^{-\Delta E/(k_B T)} \tag{4.37}$$

is assumed for temperatures well below T_N. Here ΔE is the energy barrier and α_0 is a phenomenological attempt frequency .

The ac susceptibility is measured by SQUID susceptometry (Quantum Design MPMS5S) after thermal stabilization of each temperature value of the subsequent data points. Thermal stabilization and measurement require an average time interval of $\Delta t \approx 124$ s. In the case of temperature steps of $\Delta T = 1$ K this yields the average heating rate $q = \Delta T/\Delta t \approx 8$ mK/s and a corresponding linear temporal increase of the temperature

$$T(t) = q(t - t_0) + T_s. \tag{4.38}$$

Here t_0 and T_s are the time and the temperature at the start of the measurement, respectively. In accordance with (4.38), the time dependence of the temperature affects the solution of (4.37). Integration yields

$$N(T) = N_0 e^{-\frac{1}{q} \int_{T_s}^{T} \alpha(T') dT'}, \tag{4.39}$$

the temperature dependent number of spin pairs which contribute to the total susceptibility of (4.36). Here N_0 is the total number of spin pairs which are generated during the field-cooling procedure. Equation (4.39) is related to the Randall–Wilkins equation, which is known from the physics of thermoluminescence where trapped electrons are thermally activated and give rise to light emission on heating [65, 66]. The temperature dependence of the luminescence is also known as the 'glow curve'.

In order to obtain an analytic susceptibility expression which is appropriate for a fitting procedure, the integration which enters (4.39) is approximately solved. Therefore, we expand $\alpha(T')$ into powers of $(T' - T^*)$ up to the first-order, where $T^* = (T_s + T)/2$ is the center of the integration interval. Subsequent integration with respect to T' yields

$$\int_{T_s}^{T} \alpha(T')\mathrm{d}T' \approx \alpha(T^*)\,(T - T_s) = \alpha_0 e^{-2\Delta E/(k_B(T+T_s))}\,(T - T_s). \qquad (4.40)$$

Substitution of the approximation (4.40) into (4.39) yields an explicit expression of the number of spin pairs at a given temperature T. This expression enters (4.35) via the proportionality constant C.

The horizontal domain walls (Fig. 4.15) carry an excess moment, m_c, which is expected to vanish as a consequence of domain relaxation . This is indeed observed as shown in Fig. 4.16 which exhibits the temporal relaxation, m vs. t, of the magnetic moment of a $FeCl_2/CoPt$ multilayer after field-cooling in $\mu_0 H = 5T$ to $T = 5$ K. A best fit to a stretched exponential function

$$m(t) = m_0 + m_1 e^{-(t/\tau)^\beta} \qquad (4.41)$$

yields $m_0 = 1.12 \times 10^{-7}$ Am^2 , $m_1 = 1.1 \times 10^{-8}$ Am^2, $t = 1350$ s and $\beta = 0.69$. The stretched exponential decay implies a polydispersive relaxation process. Its average relaxation time is given by $<\tau> = \tau\Gamma(1/\beta)/\beta$ [67]. With $\Gamma(1/0.69) = 0.8857$ one obtains $<\tau> = 1733$ s. This figure may be considered as the typical relaxation time of the above described metastable spin configurations, which enters the domain dynamics to be discussed below.

In close analogy to the domain-wall relaxation described above, it is reasonable to assume that one dominating activation energy, $\Delta\tilde{E}$, controls the temporal decay of the magnetic moment. This is corroborated in the inset of Fig. 4.16, which shows the 'glow curve' m vs. T and its derivative $\mathrm{d}m/\mathrm{d}T$ vs. T. The latter exhibits a pronounced minimum at $T = 15$ K, which indicates a thermally activated relaxation towards the low-moment ground state . The horizontal domain walls carry a typical magnetic moment, m_c, which build up the total surplus moment of the AF bulk. In order to rotate the spins of a cluster coherently, they have to overcome the barrier $\Delta\tilde{E}$. It is expected to depend on the applied magnetic field, ΔH, in accordance with the Zeeman energy $\Delta E = -\mu_0 m_c \Delta H$. Hence, ΔH modifies the activation energy of each cluster by ΔE which gives rise to a field-dependent relaxation rate . A positive magnetic field lowers the energy of the magnetic moments which point along the field direction. Hence, the energy barrier increases with increasing magnetic field. This causes a field-induced surplus magnetization with respect to the zero or negative magnetic field condition, where the reduction of the energy barrier accelerates the decay of the magnetization. This mechanism gives rise to a positive susceptibility contribution, χ_c , which superimposes on χ_{sp}.

The relaxation of m originates from the rearrangement of energetically unfavorably oriented spins towards the magnetized ground state . In a first approximation the flipping rate of the spins can be described by (4.37), where the new activation energy and a new attempt frequency, $\tilde{\alpha}_0$, enter $\tilde{\alpha}(T)$. The thermal evolution of the number of unfavorable oriented spins is again given by (4.39) by taking into account the new parameters $\tilde{\alpha}_0$ and $\Delta\tilde{E}$. It is

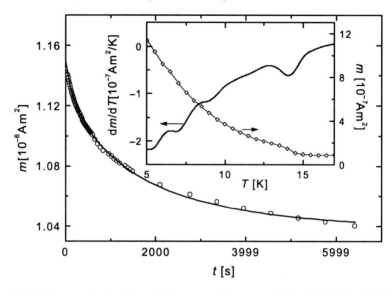

Fig. 4.16. Temporal relaxation of the magnetic moment (*circles*) at $T = 5$ K after cooling in an axial field $\mu_0 H = 5$ T. The *inset* shows the temperature dependence of the magnetic moment (*diamonds*) and its derivative (*line*) after the same field-cooling procedure and subsequent zero-field heating

transformed into an approximate analytic expression with the help of (4.40), which yields

$$m(T) = m_0 \exp\left(-\frac{\tilde{\alpha}_0(T - T_{\mathrm{s}})}{q} \mathrm{e}^{-2(\Delta\tilde{E}+\mu_0 m_{\mathrm{c}} H)/(k_{\mathrm{B}}(T+T_{\mathrm{s}}))}\right). \tag{4.42}$$

The derivative $\partial m/\partial H$ of (4.42) at $H = 0$ then yields the susceptibility contribution

$$\chi_{\mathrm{c}} = \frac{2\mu_0 m_0 m_{\mathrm{c}}\tilde{\alpha}_0(T - T_{\mathrm{s}})}{q k_{\mathrm{B}}(T + T_{\mathrm{s}})}$$

$$\times \exp\left(-\frac{\tilde{\alpha}_0(T - T_{\mathrm{s}})}{q}\mathrm{e}^{-2\Delta\tilde{E}/(k_{\mathrm{B}}(T+T_{\mathrm{s}}))} - \frac{2\Delta\tilde{E}}{k_{\mathrm{B}}(T + T_{\mathrm{s}})}\right). \tag{4.43}$$

The total susceptibility is given by the sum of the spin pair and the cluster contributions. Figure 4.17 shows the results of best fits of to the data of the excess susceptibility obtained after subtraction of the zero-field-cooled back-ground signal (Fig. 4.14).

The seven fitting parameters which enter the model to fit the experimental data of the excess susceptibility have to be discussed. The proportionality constant C, which enters the spin-pair contribution, decreases linearly:$C = 0.00144, 0.00091$ and 0.0006 for $T_0 = 11, 12$ and 13 K, respectively. It indicates that the number of wall-spins increases with decreasing

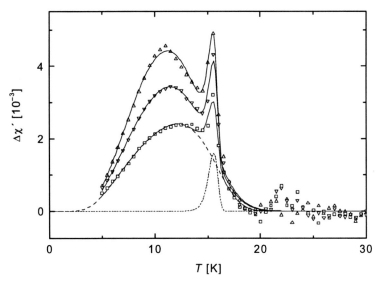

Fig. 4.17. Excess susceptibility $\Delta\chi$ vs. T obtained from Fig. 4.14 after subtraction of the zero-field-cooled curves using the same symbols as in Fig. 4.14. The *solid lines* exhibit the results of best fits of $\Delta\chi(T) = \chi_{\rm sp}(T) + \chi_{\rm c}(T)$, (4.35) and (4.43), to the data. As an example, the decomposition of $\Delta\chi$ into $\chi_{\rm sp}$ and $\chi_{\rm c}$ is shown for $T = 13$ K (*dashed lines*)

transition temperature T_0 where the antiferromagnetic ordering sets in (see the inset of Fig. 4.14). As expected, the energy parameter $\tilde{K}_2$ which enters (4.35) is virtually independent of temperature and given by $\tilde{K}_2/k_{\rm B} = 25.6$ K± 1.2 K. This is reasonable, because $\tilde{K}_2$ is determined by the microscopic exchange between the spins of a given pair and their corresponding AF ordered neighborhood. The typical interaction energy of a spin is the order $k_{\rm B}$ times the Néel temperature $T_{\rm N} = 23.7$ K of the antiferromagnet, which is in fact pretty close to the resulting fitting parameter.

In contrast to that, the energy barrier ΔE as well as the ratio of the attempt frequency α_0 and the heating rate q increase from $\Delta E/k_{\rm B} = 61.2, 79.7$ to 105.0 K and $\alpha_0/q = 44.0, 208.9$ to 1719.9 K^{-1} with increasing $T_{\rm c}(H)$, respectively. The height of the barrier is given by the energy which is required to move two adjacent walls until they meet and annihilate each other. Hence, ΔE increases with the typical domain size. The latter is expected to increase with increasing $T_{\rm c}(H)$, because small deviations from the AF ground state are efficiently quenched by thermal spin-flips which increase with increasing temperature.

The domain walls which are generated by AF stacking faults can be regarded as FM clusters. Their contributions are given by (4.43). Its fit to the data yields $\Delta\tilde{E}/k_{\rm B} = 502.5, 485.4$ and 503.7 K as well as $\tilde{\alpha}_0/q = 1.7 10 20, 2.7 10 19$ and 1.7×1020 K^{-1}, respectively. The parameters are virtually temperature-independent, indicating that the cluster size does not depend on temperature. In order to reduce the number of these domains, coherent rotation of large regions within the FM layers is necessary, which may explain the high value of the energy barrier. Moreover, in comparison with α_0 the very high values of the attempt frequency $\tilde{\alpha}_0$ indicate that the cluster excitations are given by collective modes of the spin-wave type. In contrast to the proportionality constant C, the pre-factor $P = 2\mu_0 m_0 m_c \tilde{\alpha}_0/(q k_B)$ of χ_c reveals remarkably high values of $P = 5.5 \times 10^{16}, 7.4 \times 10^{15}$ and 3×10^{16}. This is in accordance with the model assumption of AF stacking faults that carry large magnetic moments.

In contrast with the FeF_2/CoPt heterostructure where the antiferromagnet pins the FM top layer we are faced to an inverse situation when replacing FeF_2 by the soft antiferromagnet $FeCl_2$. Instead of pinning the ferromagnet, the exchange interaction at the interface creates an AF domain state which is modified during the hysteresis cycle. In detail, the CoPt layer couples to the AF spins at the uncompensated rough interface of the (111)-oriented $FeCl_2$ single crystal. The topological roughness gives rise to magnetic roughness at the interface where the growth of the AF domains sets in. The experimental results insistently demonstrate that AF domain formation plays a crucial role in exchange-bias systems where, on the one hand, pinning of the ferromagnet has to be considered, but, on the other hand, the mutual AF/FM interaction strongly influences the AF order.

4.3 Heterostructures with Planar Anisotropy and Compensated Interfaces

4.3.1 Piezomagnetic Origin of Exchange Bias

The investigation of the perpendicular exchange-bias system $FeF_2(001)$/CoPt exhibits the onset of a piezomagnetic moment on cooling the heterosystem to below T_N. In the case of $FeF_2(001)$ the piezomagnetism may be regarded as a secondary effect with minor relevance for the exchange-bias effect because the uncompensated spins at the surface of the antiferromagnet provide a net AF interface moment S_{AF}. Therefore, no additional effect is needed in order to explain the exchange-bias . However, already in the case of statistical interface roughness the condition $S_{AF} \neq 0$ requires AF domain formation [15] when piezomagnetism is neglected. Under the assumption of a perfect FeF_2 bulk crystal where defect-induced domain-wall pinning is missing, domain formation is, however, energetically unfavorable. Hence, exchange-bias is not expected a priori.

The situation becomes even more sophisticated in the case of compensated AF interfaces. In order to solve the puzzle of exchange-bias in that case it has recently been stressed that the surplus magnetic moment of random-field domains gives rise to exchange-bias in heterostructures with compensated AF interfaces [16, 18, 68]. In fact, it is well-known that a diluted antiferromagnet in a field breaks down into a random-field domain state on cooling to below T_N [23]. It carries a net magnetic moment which gives rise to an AF interface moment and thus to exchange-bias in the case of compensated AF surfaces [16].

However, in the absence of dilution exchange-bias of the same amount is also observed at a compensated AF surface [40]. In that case, the existence of an AF interface moment is usually attributed to unavoidable crystal imperfections which is, however, a less convincing assumption. Therefore, in addition to the established random-field mechanism it is worthwhile to take into account piezomagnetism as a mechanism which requires no impurities or structural defects in order to give a further possible explanation of the exchange-bias in heterostructures with compensated both diluted and pure AF pinning layers. The symmetry conditions which provide a thermodynamic potential with a linear field term suggest that only a few antiferromagnets like, e.g., α-Fe$_2$O$_3$, MnF$_2$, CoF$_2$ and, in particular, FeF$_2$, give rise to piezomagnetism [69]. However, at the AF/FM-interface significant changes of the symmetry take place with respect to the AF bulk in addition to the obvious breaking of translational invariance. Hence, the number of AF systems which develop a piezomagnetic interface moment S_{AF} is very probably much higher than the symmetry considerations of the bulk systems suggest.

In the following, the exchange-bias of Fe$_{0.6}$Zn$_{0.4}$F$_2$(110)/Fe 5 nm/Ag 35 nm and Fe$_{0.6}$Zn$_{0.4}$F$_2$(110)/Fe 14 nm/Ag 35 nm is investigated with particular attention to the impact of piezomagnetism. Both heterostructures are expected to have compensated (110)-surfaces while the shape anisotropy of the FM layer causes planar anisotropy in accordance with (3.7). Moreover, the diamagnetic dilution gives rise to prototypical random-field behavior of the antiferromagnet. Hence, large exchange-bias effects are expected in this system.

Figure 4.18 shows a sketch of the investigated heterostructrues which are grown by UHV-deposition of 5 and 14nm Fe on top of the compensated (110) surface of the diamagnetically diluted antiferromagnet which is thermally stabilized at $T = 425$ K during the growth process. In accordance with the planar anisotropy of the Fe layer, the applied magnetic field is oriented along the c axis of the antiferromagnet and lies within the easy plane of the FM-layer. The rutile structure of the bulk antiferromagnet, its corresponding spin structure and an additional top view on the (110) surface are shown in detail. The (110) orientation of the Fe$_{0.6}$Zn$_{0.4}$F$_2$ crystal has been checked by X-ray diffraction using large-angle θ-2θ scans [70]. Figure 4.19 shows the resulting diffracted intensity as a function of 2θ. The strong (110), (220) and (330) peaks indicate (110) orientation of the crystal surface. The peaks are located at $2\theta = 26.80, 55.30$ and $88.20°$ (degree), in accordance with the Bragg condition $2d_{h,k,l} \sin 2\theta = n\lambda$ where $\lambda = 1.542$Å, n is an integer number labeling the order of the Bragg reflex and $d_{h,k,l}$ is the interplanar spacing of the (hkl) planes which reads $d_{h,k,l} = 1/\sqrt{(h^2 + k^2)/a^2 + l^2/c^2}$ in the case of a tetragonal lattice . The latter formula is straightforwardly derived from the general relation $d_{h,k,l} = 2\pi/|\boldsymbol{G}_{h,k,l}|$, where the reciprocal lattice vector reads $\boldsymbol{G}_{h,k,l} = 2\pi(h/a, k/a, l/c)$ for tetragonal symmetry. While the lattice parameter $a = 4.70$ Å is identical for FeF_2 and ZnF_2, the lengths c of the tetragonal axes vary between 3.31 and 3.13 Å, respectively. However, the $(hk0)$ reflexes are not affected by the variation of c. Hence, their positions are independent of the diamagnetic dilution of the mixed crystal.

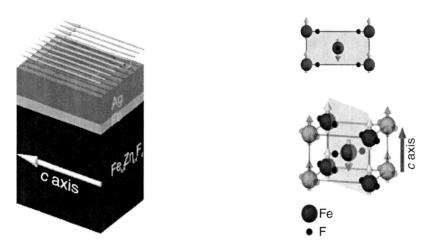

Fig. 4.18. Sketch of the planar anisotropic heterostructure $Fe_{0.6}Zn_{0.4}F_2(110)$ /Fe/Ag and the rutile structure (Fe and F ions represented by *large* and *small spheres*, respectively) of the bulk antiferromagnet. AF spin ordering is indicated by *arrows*. A top view of the (110) surface exhibits details of the compensated spin structure

In full agreement with the results of Nogués et al. [40] who investigated Fe on $FeF_2(110)$ we find a polycrystalline Fe layer with (110) and (100) texture in accordance with the strong (110), (220) and (200) Bragg peaks of Fe. In addition, a fingerprint of the Ag capping layer is given by the corresponding (111) and (222) peaks.

After field-cooling to below the Néel temperature of $Fe_{0.6}Zn_{0.4}F_2$, $T_N = 47$ K, the exchange-bias effect is investigated by SQUID magnetometry as a function of the temperature and of the magnetic moment, m_{FM}, of the FM thin-film. In contrast with the perpendicular exchange-bias system $FeF_2/CoPt$ the amount and sign of the loop shift are determined by the value and the direction of the magnetic moment of the Fe layer, m_{FM}. It is, however, independent of the freezing field apart from the fact that the latter controls the magnetic moment of the FM layer [71].

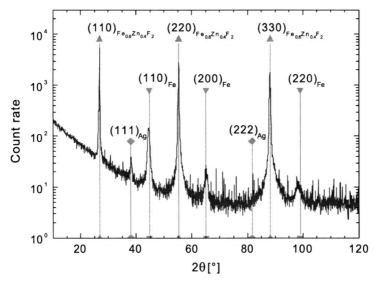

Fig. 4.19. Large-angle X-ray diffraction of $Fe_{0.6}Zn_{0.4}F_2(110)/Fe$ 14 nm/Ag 35 nm ($\lambda = 1.542$ Å) exhibiting the (110), (220) and (330) Bragg peaks of $Fe_{0.6}Zn_{0.4}F_2$ and the (110), (220) and (200) peaks of Fe as well as the (111) and (222) peaks of the Ag capping layer

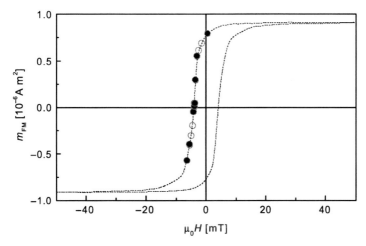

Fig. 4.20. Hysteresis loop of $Fe_{0.6}Zn_{0.4}F_2(110)/Fe$ 14 nm/Ag 35 nm obtained at $T = 100$ K by SQUID magnetometry (*line*). Starting from a saturated sample different magnetization states are prepared (*circles*) by applying various magnetic fields in the non-overshoot mode of the magnetometer in order to follow unambiguously the upper branch of the hysteresis loop from the saturation value m_s to the target value m_{FM}, where $-m_s \leq m_{FM} \leq m_s = 9.0 \times 10^{-7}$ Am^2. The loops which correspond to initial states and are indicated by *solid circles* are presented in Fig. 4.21

In order to determine m_{FM} from measurements of the total moment, the magnetic hysteresis is measured at $T = 100$ K$\approx 2T_{\mathrm{N}}$. Subsequently, a magnetic field is applied in the non-overshoot mode of the magnetometer in order to follow unambiguously the upper branch of the hysteresis loop from the saturation value m_{s} to the target value m_{FM} where $-m_{\mathrm{s}} \leq m_{\mathrm{FM}} \leq m_{\mathrm{s}} = 9.0 \times 10^{-7}$ Am^2. The magnetic field that prepares m_{FM} is applied during the freezing process. Figure 4.20 shows the hysteresis loop of Fe$_{0.6}$Zn$_{0.4}$F$_2$(110)/Fe 14 nm/Ag 35 nm at $T = 100$ K where the different magnetization states, which have been prepared before cooling, are indicated by open and solid circles, respectively. Note, that the hysteresis loop is perfectly symmetric with respect to the field and the magnetization axes in accordance with the vanishing AF order above T_{N}. Figure 4.21 shows typical hysteresis loops at $T = 10$ K obtained for freezing fields $\mu_0 H = -6.48, -4.48, -4.08, -3.98, -3.88, -3.48, -1.48$ and $+0.52$ mT which give rise to the magnetization states indicated by solid circles in Fig. 4.20. Figure 4.22 exhibits the corresponding H_{e} vs. m_{FM} dependence of the Fe$_{0.6}$Zn$_{0.4}$F$_2$(110)/Fe14nm/Ag35nm heterostructure together with a sketch of the freezing procedure. Assuming that the FM interface moment S_{FM} is proportional to the net magnetic moment of the Fe layer during the freezing process and, moreover, assuming that the AF interface moment S_{AF} is neither influenced by the exchange nor by the freezing field, the simple MB formula (4.1) predicts a linear H_{e} vs. m_{FM} dependence. In a first approximation this proportionality holds, but closer inspection shows that the data do not cross the origin of the coordinate system. Rather a small shift towards positive H_{e} values remains at $m_{\mathrm{FM}} = 0$. Moreover, on approaching $m_{\mathrm{FM}} = \pm m_{\mathrm{s}}$, H_{e} deviates from its linear m_{FM} dependence. The latter behavior indicates a weak residual dependence of S_{AF} on both the freezing and the exchange field, which arises from the AF/FM interaction at the interface. As expected, its influence increases with increasing m_{FM}. In the vicinity of $m_{\mathrm{FM}} = 0$, however, it is reasonable to assume $S_{\mathrm{AF}} \approx const.$ As will be shown below, the shift of the H_{e} vs. m_{FM} curve agrees with the MB model, when generalizing the approach to a non-uniformly magnetized ferromagnet. In accordance with [71], $\pm m_{\mathrm{s}}$ yields $\mp H_{\mathrm{e}}$.

In contrast to the perpendicular exchange-bias system FeF$_2$(001)/CoPt where a strong freezing field dependence above the saturation field of the FM layer has been observed (Fig. 4.10) here the exchange-bias is controlled by m_{FM}. As pointed out above, this behavior indicates that the AF interface moment S_{AF} is virtually unaffected by the applied magnetic field. This may originate from the dominating influence of the exchange field which is expected to be much stronger than the applied magnetic field. However, in the case of a rough AF/FM interface the same argument should apply for the perpendicular exchange-bias system, which exhibits a strong freezing field dependence. In order to overcome this discrepancy a positive AF/FM coupling has to be encountered which prevents a competition between the exchange

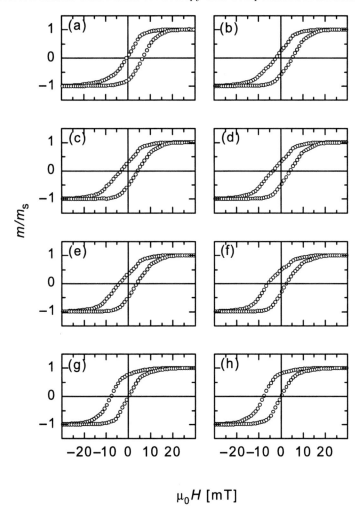

$\mu_0 H$ [mT]

Fig. 4.21. Typical hysteresis loops obtained at $T = 10$ K for freezing fields $\mu_0 H =$ -6.48 (**a**), -4.48 (**b**), -4.08 (**c**), -3.98 (**d**), -3.88 (**e**), -3.48 (**f**), -1.48 (**g**) and +0.52 mT (**h**) which give rise to the magnetization states $-m_s \leq m_{FM} \leq m_s = 9.0 \, 10^{-7}$ A m^2 indicated by *solid circles* in Fig. 4.20

and the Zeeman energy. Alternatively, however, a dominating and freezing field-independent contribution to S_{AF} like the piezomagnetic moment is able to explain the independence of H_e on freezing fields $H_F > H_s$, the saturation field of the FM layer. Nevertheless, in both cases it is reasonable to start from the ansatz

$$H_e = \left| H_e^+ \right| a - \left| H_e^- \right| (1 - a). \tag{4.44}$$

Here a is the relative portion of the total area of the Fe layer where the local magnetization is negative and, hence, the local exchange-bias field is

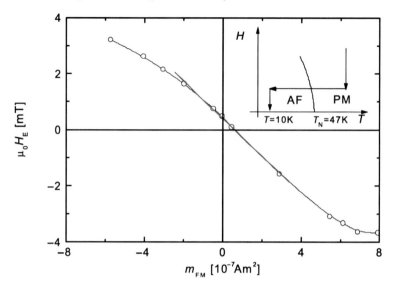

Fig. 4.22. H_e vs. m_{FM} dependence of the $Fe_{0.6}Zn_{0.4}F_2(110)/Fe$ 14 nm/Ag 35 nm heterostructure together with a sketch of the freezing procedure

given by $H_e^+ > 0$, while $(1-a)$ is the remaining part, where the magnetization is positive and $H_e^- < 0$. The shift of the total measured hysteresis is given by the sum of the local contributions weighted with respect to the relative areas. In accordance with the conventional MB formula, H_e^+ and H_e^- are controlled by the local interface moments. This reads

$$H_e \propto \left|S_{FM}^+ S_{AF}^+\right| a - \left|S_{FM}^- S_{AF}^-\right| (1 - a). \tag{4.45}$$

It is reasonable to assume that the magnitude of the FM interface moment per unit area does not depend on the sign of the local magnetization, $\left|S_{FM}^-\right| = \left|S_{FM}^+\right| = \left|S_{FM}\right|$. However, the AF interface moment per unit area may depend on the orientation of the local FM interface moment $\left|S_{AF}^-\right| = \left|S_{AF}^+\right| - \Delta S_{AF}$. A microscopic justification of a finite deviation, $\Delta S_{AF} \neq 0$, is given below.

Substitution of $\left|S_{AF}^-\right|$ into (4.45) yields

$$H_e \propto \left|S_{FM}\right| \left((2\left|S_{AF}^+\right| - \Delta S_{AF})a - \left|S_{AF}^+\right| + \Delta S_{AF} \right). \tag{4.46}$$

The total magnetic moment of the Fe layer is given by $m_{FM} = -m_s a + m_s(1-a)$, the sum of domain contributions with positive and negative saturation magnetization. Hence, the normalized area can be expressed according to $a = (1 - m_{FM}/m_s)/2$. Substitution of a into (4.46) yields the explicit m_{FM} dependence of H_e. It reads

$$H_e \propto \left|S_{FM}\right| \left(\frac{-(2\left|S_{AF}^+\right| - \Delta S_{AF})m_{FM}}{2m_s} + \frac{\Delta S_{AF}}{2} \right). \tag{4.47}$$

Obviously, (4.47) describes the observed shift of H_e vs. m_{FM} (Fig. 4.22) in the case $\Delta S_{AF} \neq 0$.

Whenever the AF bulk breaks down into a domain state on cooling in a freezing field to below T_N, the magnitude and orientation of the AF interface moments are controlled by the competition between the exchange interaction with the adjacent FM layer and the adaptation of the interface spin configuration to the underlying AF domain structure. Only in the case of very strong exchange interaction at the interface will the ferromagnet completely control the orientation of the AF interface moment so that $S_{AF}^+ = -S_{AF}^-$. However, in the case of a 'strong' antiferromagnet like $Fe_{0.6}Zn_{0.4}F_2$ a compromise between complete interface and bulk adaptation, respectively, has to be found. Hence, $\Delta S_{AF} \neq 0$ has to be expected in the case of AF domain states, which do not match perfectly with the FM ones.

Although random field provide an AF domain state, here an alternative mechanism is proposed which originates from piezomagnetism . It has the advantage to explain a uniform shift of the H_e vs. m_{FM} curve. In the case of a random-field domain state it should depend on the freezing field and, in particular, on the strong exchange field. The latter is expected to decrease with decreasing $|m_{FM}|$, if the FM domain size is smaller than the size of the AF domains.

As pointed out in Sect. 4.2.2, vertical shifts of the hysteresis curves of $FeF_2(110)/Fe$ originate from the piezomagnetic moment. The same situation holds in the diluted heterostructure $Fe_{0.6}Zn_{0.4}F_2(110)/Fe$ 5 nm/Ag 35 nm, where the onset of a piezomagnetic moment on cooling to below T_N is demonstrated by the of m vs. T behavior.

Figure 4.23 shows the m vs. T data measured by SQUID magnetometry with (squares) and without (circles) external shear-stress $\sigma_{xy} > 0$ applied along the [110] direction of the antiferromagnet. The latter modifies the piezomagnetic moment [50]

$$m_z^P = \lambda \sigma_{xy} l_z / |\underline{l}|, \tag{4.48}$$

by changing the natural stress distribution $\sigma_{xy}(\boldsymbol{r})$. Under an applied freezing field, the evolving piezomagnetic moment, m_z^P , will minimize its Zeeman energy . Hence, a built-in shear-stress distribution $\sigma_{xy}(\boldsymbol{r})$ with changing signs gives rise to changing signs of the AF vector l and its z component l_z. Obviously, piezomagnetism creates an AF state which carries a magnetic moment and breaks down into domains. Its interface contribution affects the exchange-bias field in accordance with the MB approach. In addition, the domain formation will give rise to $\Delta S_{AF} \neq 0$.

Figure 4.24 shows the magnetic hysteresis after cooling from $T = 100$ K to 10 K in a freezing field of $\mu_0 H = 5$ mT with (squares) and without (circles) external shear-stress σ_{xy}. The [110] oriented stress originates from two copper plates which apply pressure on the top and bottom surfaces of the heterostructure. Copper wires which shrink on cooling connect the upper and lower plates and generate shear-stress which reduces the piezomagnetic

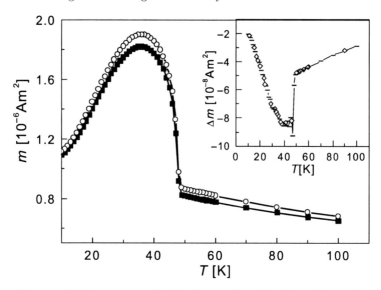

Fig. 4.23. m vs. T data measured with (*squares*) and without (*circles*) external shear-stress $\sigma_{xy} > 0$ applied along the [110] direction of the antiferromagnet. The *inset* exhibits the difference (*diamonds*) between both sets of data

moment (see inset of Fig. 4.24). In accordance with the behavior of m vs. T (Fig. 4.23) the reduced shift of the hysteresis indicates that the built-in stress has a negative sign on the average. Hence, in accordance with the MB model, the magnitude of H_e decreases from 25.3 mT (Fig. 4.24 circles) to 23.1 mT (Fig. 4.24, squares) on applying external shear-stress.

Although the piezomagnetism in $Fe_{1-x}Zn_xF_2$ is a well-known bulk phenomenon [51] and evidenced for the heterosystem by the above findings, its contribution to the AF interface moment is not quantitatively determined so far. Hence, presently it remains an open question whether piezomagnetism alone can be the origin of exchange-bias. However, as outlined above there are several indications that piezomagnetism plays a crucial role for exchange-bias.

One additional piece of evidence is the fact that the training effect in exchange-bias systems based on AF bulk or thin-film single crystals is very small or even non-existent [43, 10, 21, 40]. The reduction of natural built-in stress is effectively suppressed in single crystals and, hence, piezomagnetically induced exchange-bias becomes possible. In exchange-bias systems based on polycrystalline antiferromagnets the crystallites have the possibility to relax and, hence, to reduce the built-in stress. Therefore, in the polycrystalline antiferromagnets S_{AF} depends on AF domain formation only. The rearrangement of the corresponding spin configurations, which has been predicted by Monte Carlo simulations [18], gives rise to training effects where the exchange-bias decreases typically according to $H_e - H_{e_\infty} \propto 1/\sqrt{n}$ with an increasing num-

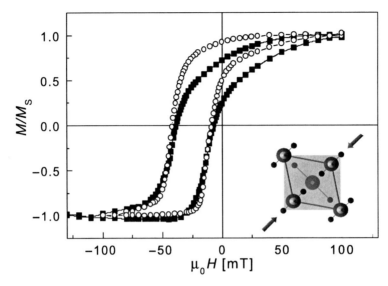

Fig. 4.24. Magnetic hysteresis after saturation of the FM layer at $T = 100$ K and subsequent cooling in $\mu_0 H = 5$ mT to $T = 10$ K. Circles indicate the hysteresis under natural shear-stress. Squares exhibit the hysteresis on applying external shear-stress $\sigma_{xy} > 0$. The inset exhibits the influence of $\sigma_{xy} > 0$, which destroys the equivalence of the two AF sublattices

ber n of loops [18, 21, 22]. A more detailed discussion of the training effect is given in Sect. 4.3.3.

4.3.2 Dilution Controlled Temperature Dependence of Exchange Bias

The temperature dependence of the exchange-bias field of the planar heterosystem $Fe_{0.6}Zn_{0.4}F_2(110)/Fe/Ag$ deviates significantly from the H_e vs. T dependence of the perpendicular exchange-bias system $FeF_2(001)/Fe/Ag$ (Fig. 4.12). Figure 4.25 shows the H_e vs. T -data (circles) of $Fe_{0.6}Zn_{0.4}F_2(110)$ /Fe 14 nm/Ag 15 nm. The data are obtained from magnetic hysteresis curves at the respective target temperatures $5\,K \leq T \leq 80\,K$ after cooling from $T = 100$ K in a freezing field of 5 mT which has been applied parallel to the planar [110] direction. A typical hysteresis curve at $T = 5$ K is shown in the inset of Fig. 4.25. In contrast with the perpendicular exchange-bias system which shows a virtually linear temperature dependence of H_e, here the temperature dependence of the data suggests a typical order-parameter behavior where H_e vs. T increases with decreasing temperature closely following a power law.

As pointed out in the last section the maximum exchange-bias field of $Fe_{0.6}Zn_{0.4}F_2(110)/Fe/Ag$ occurs on cooling the system to below T_N while the magnetization of the *FM* Fe layer is completely saturated. No significant change of H_e is observed when increasing the freezing field up to 5 T far above the saturation field. This behavior suggests positive exchange coupling between S_{FM} and S_{AF}. This situation excludes competition between the Zeeman energy of the AF interface spins and the AF/FM exchange during the freezing process. Owing to the analysis presented in Sect. 4.2.2 such a competition favors the temperature-activated rearrangement of the AF spins at the interface and gives rise to vanishing exchange-bias significantly below T_N.

However, in the case of FM coupling at the (110) surface the temperature dependence of H_e depends on the thermal evolution of the AF surface order-parameter $\eta_{AF}^s \propto |T - T_N|^{\beta_s}$. It is reasonable to assume that η_{AF}^s is proportional to S_{AF} which determines the temperature dependence of H_e in accordance with (4.1). In the case of a piezomagnetic origin of exchange-bias the proportionality between the AF order-parameter and the piezomagnetic moment is described by (4.48) and, moreover, has been experimentally verified by Mattsson et al. [72].

Under the assumption $\eta_{AF}^s \propto S_{AF}$ (4.1) suggests that the exchange-bias effect vanishes at the Néel temperature of the bulk antiferromagnet .

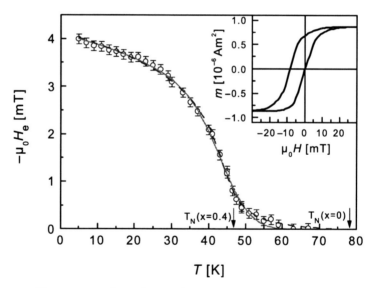

Fig. 4.25. Temperature dependence of the exchange-bias field $\mu_0 H_e$ (*circles*) in $Fe_{0.6}Zn_{0.4}F_2(110)$ /Fe 14 nm/Ag 15 nm obtained from magnetic hysteresis curves at the respective target temperatures $5\,K \leq T \leq 80\,K$ after cooling in a freezing field of 5 mT. The *inset* shows a typical hysteresis curve measured at $T = 5$ K. The solid and *dashed lines* show the results of the best fits of (4.49) involving Gaussian and Lorentzian distribution functions, respectively. *Arrows* at 46.9 and 78.4 K indicate the Néel temperatures of the diluted and the pure antiferromagnet, respectively

Note, that in general the Curie temperature of the ferromagnet is much higher than T_N and has no influence on the temperature dependence of H_e. A rare exception has been reported in [14, 73], where FM/AF bilayers with comparable critical temperatures T_c and T_N exhibit an enhanced exchange-bias field on approaching $T_c \gtrsim T_N$. Typically, the blocking temperature, T_B, i.e. the temperature of vanishing exchange-bias effect, is smaller than T_N [21]. This behavior is known from our perpendicular exchange-bias system (see Fig. 4.12) and, moreover, has been found in various planar systems. In the latter it is usually attributed to finite size effects, which are known, e.g., from the dependence of the EB on the AF layer thickness [17, 18, 19].

Lederman et al. [56] showed that in FeF_2–Fe bilayers near T_N the temperature dependence of the exchange-bias field follows the temperature dependence of the AF interface magnetic moment which exhibits power-law behavior owing to surface criticality. In the case of large grain sizes ξ_g, the corresponding critical exponent $\beta_s = 0.8 \pm 0.04$ agrees with the surface critical exponent of the 3d Ising system. With decreasing grain sizes ξ_g, however, the corresponding decrease of β_s was attributed [56] to either a finite size effect in accordance with an effective decrease in lateral terrace size or to an increase of the interface interaction. The latter gives rise to surface ordering which occurs independently of the bulk. Moreover, an increased interface interaction may explain a blocking temperature slightly above T_N.

However, in accordance with Fig. 4.25 the blocking temperature $T_B \approx 63$ K of $Fe_{0.6}Zn_{0.4}F_2(110)$/Fe 14 nm/Ag 35 nm is tremendously enhanced with respect to the Néel temperature $T_N = 46.9$ K of the diluted bulk antiferromagnet $Fe_{0.6}Zn_{0.4}F_2$. The latter is determined from the temperature dependence of the magnetic moment of the heterosystem shown in Fig. 4.26. The Curie temperature of Fe is much higher than the global T_N of $Fe_{0.6}Zn_{0.4}F_2$. Hence, the magnetic moment of Fe remains almost constant at all temperatures under investigation and the m vs. T dependence mainly reflects the temperature dependence of the parallel magnetic susceptibility of the antiferromagnet. However, the steep increase in the vicinity of T_N on the one hand and the delayed decay of the magnetic moment above T_N originate very probably from the residual interface interaction. The inset of Fig. 4.26 exhibits the first and the second derivatives of $m(T)$, which indicate the Néel temperature $T_N = 46.9$ K of the $Fe_{0.6}Zn_{0.4}F_2$ substrate.

The huge enhancement of T_B with respect to T_N cannot be explained within the framework of a small proximity effect that originates from the mutual AF/FM interaction at the interface. Instead we propose that clusters with arbitrary concentration of ZnF_2, $0 \leq x \leq 1$, originate from fluctuations of the diamagnetic dilution occurring in the natural growth of the AF bulk crystal. They appear on all length scales, where infinite clusters give rise to the well-known Griffiths phase implying non-analyticity within $T_N \leq T \leq T_N(x = 0) = 78.4$ K as predicted theoretically [74]. However, already finite clusters which are far more frequent than the exponentially rare clusters of

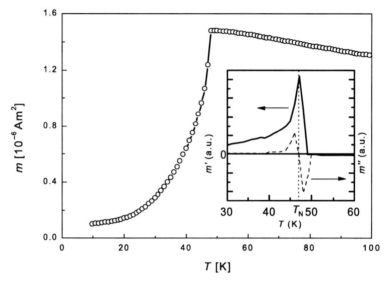

Fig. 4.26. Temperature dependence of the magnetic moment m of $Fe_{0.6}Zn_{0.4}F_2(110)/Fe$ 14 nm/Ag 35 nm. The *inset* shows the first (*solid line*) and second (*broken line*) derivatives with respect to temperature, thus defining the Néel temperature of the $Fe_{0.6}Zn_{0.4}F_2(110)$ substrate, $T_N = 46.9$ K

infinite size are expected to provoke a local enhancement of the exchange-bias. In the following, an analysis of the temperature dependence of the integral exchange-bias effect is presented.

To this end the local contributions to the AF surface magnetization, which exhibit critical behavior, are averaged with respect to a phenomenological cluster distribution under the assumption of local quasi-critical temperatures.

Lederman et al. [56] pointed out that in the FeF_2–Fe system the temperature dependence of H_e should be directly proportional to the AF surface order-parameter. In order to take into account the influence exerted by perturbations like strain and disorder on the exchange-bias effect they introduced the "rounded" power law

$$H_e(T) = H_e^0 \int_0^\infty t^{\beta_s} P(T_c) dT_c, \qquad (4.49)$$

where $t = 1 - T/T_c$ is the reduced temperature for $T < T_c$ and $t = 0$ for $T > T_c$ while $P(T_c)$ is the critical temperature distribution. A narrow Gaussian distribution function,

$$P(T_c) = \frac{1}{\Delta\sqrt{2\pi}} \exp[-(T_c - T_{C0})^2/2\Delta^2], \qquad (4.50)$$

has been used for a best fit of the H_e vs. T data, where H_e^0, T_{C0}, β_s and Δ are the fitting parameters.

In contrast with the assumption of weak perturbations in FeF$_2$, however, here the fact is stressed that the diamagnetic dilution is a strong and intrinsic perturbation. It gives rise to a broad distribution function even in the case of an ideal antiferromagnet with random site dilution .

Previously it has been shown that the dilution-induced Griffiths phase of the antiferromagnet $Fe_{1-x}Zn_xF_2$ gives rise to significant deviations from the Curie–Weiss behavior which is naively expected at $T > T_N$ [75, 76]. However, Griffiths-type clusters give rise to a continuous series of local phase transitions with quasi-critical temperatures T_c within $T_N(x) \leq T_c \leq T_N(x = 0)$.

In view of this experience, (4.49) is fitted to the experimental data of Fig. 4.25 (circles) while using alternatively two phenomenological distribution functions, viz. the Gaussian distribution (4.50) and the Lorentzian distribution

$$P(T_c) = \begin{cases} \dfrac{\epsilon/2}{[\epsilon^2+(T_c-T_{C0})^2]\arctan[(T_N(0)-T_{C0})/\epsilon]} \quad) \; if \; T_c \leq T_N(x = 0), \\ \\ \qquad\qquad 0 \qquad\qquad\qquad if \; T_c > T_N(x = 0). \end{cases} \quad (4.51)$$

Here $T_N(0) = 78.4$ K is the Néel temperature of pure FeF$_2$ while H_e^0, T_{C0}, β, Δ and ϵ the fitting parameters. The Lorentzian distribution function has been used previously in order to model the T_c-distributon involved in the analogous phenomenon of a field-induced Griffiths phase [74]. It is normalized under the constraint $\int_{T_{c0}}^{T_N(0)} P(T_c)dT_c = 1/2$, where a symmetric distribution with respect to T_{C0} is assumed. T_{C0} describes the centers of gravity of the distribution functions, respectively, and is expected to be in the vicinity of the global transition temperature $T_N(x) = 46.9$ K.

Figure 4.25 shows the results of the best fits of (4.49) involving the Lorentzian distribution (dashed line) and the Gaussian distribution (solid line), respectively. The resulting fitting parameters are listed in Table 4.1 and subsequently discussed.

Table 4.1. List of the parameters resulting from the best fits of (4.49) to the $\mu_0 H_e$ vs. T data (Fig. 4.25, *circles*) involving Gaussian and Lorentzian distribution functions, respectively

	$\mu_0 H_e^0 \, [mT]$	T_{C0}	β_s	$\Delta \, or \, \epsilon \, [K]$
Gaussian distribution	4	45.0	0.22	6.1
Lorentzian distribution	4.1	44.4	0.19	4.0

Inspection of (4.49) exhibits the physical meaning of $\mu_0 H_e^0$ which is the zero-temperature limit of the exchange-bias field. Hence, $\mu_0 H_e \approx 4$ mT is in accordance with the low temperature value of the $\mu_0 H_e$ vs. T data shown in Fig. 4.25. As expected, $T_{C0} = 45$ and 44.4 K, respectively, are close to the global transition temperature $T_N(x = 0.4) = 46.9$ K. The parameters $\delta = 6.1$ and $\epsilon = 4.0$ characterize the widths of the Gaussian and the Lorentzian distribution functions, respectively. Physically the most interesting parameters are the surface critical exponents $\beta_s = 0.22$ and 0.19 for the Gaussian and the Lorentzian distribution, respectively. These values are surprisingly low in comparison to $\beta_s = 0.8$, the surface critical exponent of the 3d Ising system which has been found in the FeF$_2$–Fe bilayer system [56]. However, it is known that surface critical behavior depends on the ratio between the surface coupling J_s and the bulk coupling constant J_b, $r = J_s/J_b$ [56, 77, 78, 79]. The phase diagram of the semi-infinite Ising model exhibits a critical ratio at $r_c \approx 1.50$ [77, 80]. In the case $r < r_c$ an 'ordinary transition' occurs, which depends on the bulk critical behavior. At $r > r_c$ a surface transition occurs which is independent of the bulk transition, while a so-called 'special transition' occurs at the multicritical point r_c [78].

In the above case, β_s is quite close to the value of the special transition where $\beta_s = 0.237$ and 0.26 have been reported in [80] and [81], respectively. As pointed out in the work of Lederman et al., β_s decreases with increasing interface exchange coupling because the surface spins tend to order independently of the bulk, i.e. with an exponent closer to the 2d Ising value $\beta_s = 1/8$ [82]. In the case of the diluted antiferromagnet this tendency may be enhanced, because the diamagnetic dilution gives rise to broken magnetic 'bonds'. Hence, the interaction of an AF surface moment with neighbors from the AF bulk is effectively reduced with respect to the exchange interaction between the AF and FM moments at the interface. Therefore the ratio of the effective surface and bulk interaction constants increases, thus giving rise to a reduced β_s.

4.3.3 Training Effect

FM hysteresis and, in particular, hysteretic reversal of the magnetization in exchange-bias heterostructures are both non-equilibrium phenomena from a thermodynamic point of view. Exchange bias in the case of compensated AF interfaces requires a mechanism which creates a net magnetization of the antiferromagnet, in particular at the AF/FM interface, which gives rise to exchange coupling with the ferromagnet. Although mechanisms of domain formation like random-fields [23, 16, 83] or piezomagnetism [61, 51, 50] depend on the particular AF system, the AF domain state of a three-dimensional system is, as a rule, metastable. This gives rise to non-equilibrium phenomena in the temperature- or field-driven change of magnetization. Hence, it is not surprising that the exchange-bias shows a training effect as a consequence of metastability. An extreme case of this type of metastability has been discussed in Sect. 4.2.3 where the soft antiferromagnet FeCl$_2$ is no longer able to pin the FM top layer. Instead, the magnetization-reversal of the ferromagnet rearranges the AF domain state of the adjacent antiferromagnet from the interface up to the bulk. Here, the nature of the interface exchange as a mutual interaction becomes obvious.

The training effect describes the decrease of the exchange-bias field when cycling the system through several consecutive hysteresis loops. Let n be the number of such hysteresis loops. One often finds that the exchange-bias field after n loops, H_e^n, can be described by the proportionality $H_e^n - H_e^\infty \propto \frac{1}{\sqrt{n}}$ [84, 21]. The training effect in general has its origin in the reorientation of AF domains at the AF/FM interface which takes place during each magnetization-reversal of the FM top layer [83]. As pointed out by Nogués and Schuller [21] a pronounced training effect has been found in heterosystems involving polycrystalline AF pinning layers [85, 86, 87], while in single-crystalline pinning systems this effect is expected to be small. The grain size of a polycrystalline AF pinning substrate is an upper bound for the correlation length of the AF order-parameter. Hence, polycrystallinity limits the long-range AF order and favors a metastable domain configuration. However, there are various other mechanisms, like structural disorder at the interface or impurity induced random-fields, which give rise to AF domain formation and, hence, make a training effect possible in heterostructures involving single-crystalline AF pinning layers.

AF domain states have been extensively studied in the case of NiO [88, 89, 90, 68, 91, 92, 93]. Figure 4.27 shows the variety of AF domains which will establish on cooling NiO to below its Néel temperature. They grow as a consequence of simultaneous nucleation of the AF phase at various positions of the sample while the registration of the sublattices at each nucleation center is random. Hence the AF domain state is typically not the ground state of the system which is conventionally long-range-ordered. Note that this kinetic effect for AF domain formation is very much different from FM domain growth. In the latter case, the deviation from the thermodynamic ground states give rise to forces which drive the system into the domain state. Neglecting the elastic properties of the FM material its ground state is determined by the minimization of the free-energy. It is essentially given by the competing stray-field and domain-wall energies, respectively [94].

Indeed, a distinct training effect is observed when depositing a thin Fe layer on top of the compensated (001) surface of a NiO single crystal.

As discussed in Sect. 4.1 the presence of AF interface magnetization S_{AF} is crucial for the exchange-bias. Moreover, the training effect is expected to originate from the decrease of S_{AF} with increasing n, because all other quantities which enter (4.1) are expected to be independent of n. It is therefore the aim to evidence that the training effect originates from the n-dependence of S_{AF}. For this reason we precisely determine the saturation value of the magnetization of the total heterostructure and investigate its cycle dependence. The total saturation magnetization contains a constant as well as a cycle-dependent contribution. We show that the latter exhibits the same n-

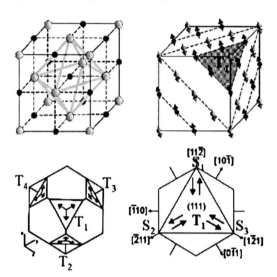

Fig. 4.27. The *upper left picture* shows the sodium chloride structure of NiO together with the ionic environment of a Ni^{2+} ion which determines the crystal field. The *upper right picture* shows the antiferromagnetic order of NiO in a T_1 single-domain state. The lower left picture illustrates the four different types of T domains which correspond to the four equivalent $< 111 >$ directions. The *lower right picture* shows the three possible S domains within the T_1 domain

dependence as the exchange-bias field and, hence, gives rise to the training effect.

A thin Fe layer of 12 nm thickness has been deposited under UHV conditions on top of the compensated (001) surface of a NiO single crystal. Its surface has been cleaned ex-situ by ion beam sputtering.

The NiO crystal was thermally stabilized at $T = 142$ K during Fe-evaporation at a rate of 0.01 nm/s. An additional double cap layer of 3.4 nm Ag and 50 nm Pt prevents oxidation of the Fe film and therefore allows ex-situ characterization. Figure 4.28 a shows the result of large-angle X-ray diffraction using Cu_{K_α} radiation. It exhibits (110)-texturing of the Fe layer in accordance with the (110) and (220) Bragg peaks at $2\Theta_{110_{Fe}} = 44.7°$ and $2\Theta_{220_{Fe}} = 65.1°$ respectively. The (002) and (004) peaks at $2\Theta_{002_{NiO}} = 43.2°$ and $2\Theta_{004_{NiO}} = 94.8°$ evidence the (001) orientation of the surface of the NiO single crystal in accordance with previous X-ray analyses [38, 39].

After growing, the sample was cooled in a planar magnetic field of 0.5T from 673 K to room temperature under high-vacuum condition. This field-cooling procedure induces the unidirectional magnetic anisotropy, which gives rise to the exchange-bias. Starting from the virgin state of the heterostructure we subsequently measured magnetization hysteresis loops in a planar applied field at $T = 5$ K using SQUID magnetometry.

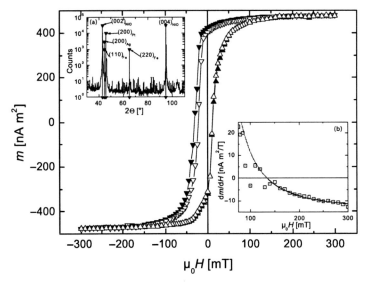

Fig. 4.28. Magnetic hysteresis curves of the NiO/Fe heterostructure after subtraction of a linear background contribution. The *solid* and *open triangles* show the first and the ninth hysteresis loops of the virgin sample after field-cooling. *Up* and *down triangles* indicate the up and down branches of the hysteresis, respectively. *Inset* (**a**) displays the X-ray-diffraction pattern with *vertical lines* indicating the Bragg peaks. *Inset* (**b**) shows the derivative $dm/d\mu_0 H$ (*dot-centered squares*) of the upper part of the down branch of one exemplary hysteresis curve. The *bold solid line* represents the fit (see text) to the corresponding data. The *thin dotted line* is an extrapolation of this fit to data outside of the fitting interval

The first loop exhibits an exchange-bias of 9.51 mT. It decreases with increasing n down to 5.43 mT for $n = 9$. Figure 4.28 shows the first and ninth hysteresis loops (solid and open triangles, respectively) after subtraction of a small diamagnetic background contribution. The background subtraction turns out to be crucial for the determination of the absolute saturation values. Hence, we describe the procedure in more detail. The total moment m which is measured by SQUID magnetometry reads

$$m = m_{\mathrm{FM}} + m_{\mathrm{AF}} + m_{\mathrm{D}}, \tag{4.52}$$

where m_{FM}, m_{AF} and m_{D} are the contributions of the FM Fe layer, the AF NiO single crystal and a diamagnetic background contribution of the sample holder. The last is known to be an odd function of the magnetic field, which originates from a field-independent susceptibility, $\chi_{\mathrm{D}} < 0$. The n-dependence of m originates from the AF contribution, m_{AF}. On the one hand, m_{AF} contains the domain induced contribution which gives rise to the training effect. On the other hand, the AF susceptibility $\chi_{\mathrm{AF}} > 0$ gives rise to a field-induced, but reversible component. Close to saturation, where nucleation processes in the magnetization-reversal of the Fe layer are negligible, the FM hysteresis can be described by the mean-field approach [95]

$$m_{\mathrm{FM}} = m_{\mathrm{FM}}^0 L \left(\frac{g\mu_B\mu_0}{k_B T} \left(H + \lambda m_{\mathrm{FM}} \right) \right), \tag{4.53}$$

where m_{FM}^0 is the saturation moment and $L(x)$ is the Langevin function. In this field regime in good approximation λm_{FM} is constant with respect to the variation of H which yields $\lambda m_{\mathrm{FM}} \approx \lambda m_{\mathrm{FM}}^0$. Using the derivative of (4.53) plus the additional constant $\chi = \chi_D + \chi_{AF}$, we fit the $dm/d(\mu_0 H)$ vs. $\mu_0 H$ data of each of the nine hysteresis loops, respectively. As a result we obtain $\chi \approx -1.34 \times 10^{-8}$ A m^2/T from the arithmetic average of the nine particular results. A typical fit is shown in the inset of Fig. 4.28 (solid line). Only data close to saturation, where $dm/d(\mu_0 H) \tilde{<} 0$, are involved in the fit. After subtraction of the background signal the exchange-field is calculated according to $H_e = \frac{1}{2}(H_{C_1} + H_{C_2})$. The coercive fields H_{C_1} and H_{C_2} are determined from the intersections of the m vs. H data with the field axis.

Figure 4.29 shows the exchange-bias field and the coercive fields ((a) and (b)) as functions of n. Obviously the training effect originates from the n-dependence of H_{C_1} while H_{C_2} shows a qualitatively similar but by far less pronounced dependence on n.

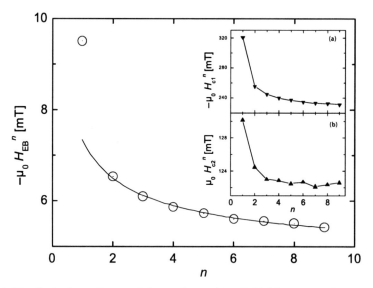

Fig. 4.29. Cycle dependence of the exchange-bias field (*dot-centered circles*) and fit (*solid line*) of (4.54) to the data for $n > 1$. The *insets* (**a**) and (**b**) show the cycle dependence of the coercive fields H_{C_1} (*down triangles*) and H_{C_2} (*up triangles*), respectively

The quantitative different n-dependences of H_{C_1} and H_{C_2} indicate that the down and up branches of the hysteresises follow different mechanisms of magnetization-reversal. A similar behavior was found on CoO/Co bilayers where coherent rotation has been observed at H_{C_2} while domain nucleation and wall propagation dominates in the vicinity of H_{C_1} [96]. Moreover, in this case it was found that the first magnetization-reversal in the virgin state is exceptional in the sense that pure $180°$ domain-wall movement takes place at H_{C_1}. In accordance with these findings we observe in our NiO/Fe heterostructure an exceptional large training effect between the first and the second hysteresis loops. As reported in very early work [84] a similar behavior has been observed for various exchange-bias systems. The best-fitted solid line in Fig. 4.29 shows that all subsequent data points $n \geq 2$ of $-\mu_0 H_e$ vs. n follow nicely the proportionality

$$H_e^n - H_e^\infty \propto \frac{1}{\sqrt{n}}. \tag{4.54}$$

The data point at $n = 1$ significantly exceeds the value when extrapolating the fit to $n = 1$ in accordance with previous investigations [84, 15].

In order to evidence the correlation between the training effect for $n > 1$ and the decrease of m_{AF} with increasing n we calculate the total saturation magnetization of the heterostructure after subtraction of χH.

The smallness of the n-dependence of the total moment makes it necessary to take advantage of averaging the m vs. H data of the n th down branch $m_n(H)$ according to

$$\bar{m}_n = \frac{1}{\Delta H_2} \int_{H_{c1}+\Delta H_1}^{H_{c1}+\Delta H_1+\Delta H_2} m_n(H) \mathrm{d}H, \tag{4.55}$$

where $\Delta H_1 = 149.5$ mT and $\Delta H_2 = 150.5$ mT. Note that it is crucial to subtract the background before calculating $\overline{m_n}$, because the interval of integration shifts proportionally with H_{C_1}. The shift of the interval of integration as a function of H_{C_1} is of major importance in order to make sure that the integration takes place in the same quasi-saturated FM state for each of the nine analyzed branches. Doing so, the above procedure prevents an artificial correlation between $\overline{m_n}$ and H_e.

The inset of Fig. 4.30 shows the $\overline{m_n} - \overline{m_{n=9}}$ vs. n data together with an eye-guiding line. By this procedure the large, but constant FM saturation magnetization, $m_{FM} \approx 10^3 \, (\overline{m_1} - \overline{m_9})$ is eliminated. Obviously, the decrease of $\overline{m_n} - \overline{m_9}$ vs. n shows a qualitative similarity with the n-dependence of the exchange-bias field in Fig. 4.29. In order to evidence direct proportionality between H_e^n we plot $-\mu_0 H_e^n$ vs. $\overline{m_n} - \overline{m_9}$ in Fig. 4.30. Note that the values $\overline{m_1} - \overline{m_9}$ come close to the limit of the SQUID sensitivity of approximately 100 p Am^2. Hence, it is necessary to determine $\overline{m_n}$ from the statistical method outlined above. The straight line in Fig. 4.30 shows the result of a best linear fit. Within the accuracy of the measurement the data follow the linear behavior apart from the first point, which is known to reflect its own rule. The data evidence that the training effect is correlated with a reduction of the AF moment. However, the integral SQUID measurement cannot distinguish between the AF bulk and interface moments. Therefore the results allow two alternative interpretations.

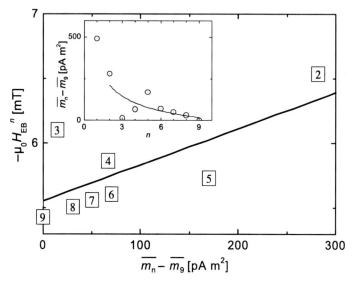

Fig. 4.30. Correlation between the exchange-bias value H_e^n and the excess moment $\overline{m_n} - \overline{m_9}$ of the sample, where the *small numbers inside the squares* indicate n. The *inset* shows the cycle dependence of the excess moment (*dot-centered circles*), a best-fitted line (*solid line*) and its extrapolation to cycle $n = 1$, which is not involved in the fit

On the one hand the AF bulk and the AF interface moments might decrease by the same ratio for each cycle. On the other hand, a constant AF bulk moment with an AF interface moment that follows the n-dependence of the exchange-bias also gives rise to a linear dependence of the exchange-bias field on the total magnetic moment. Despite this ambiguity a linear dependence of the exchange-bias field on the AF interface moment is crucial in order to explain our experimental data. Hence, the simple Meiklejohn Bean approach according to (4.1) which points out a linear dependence of H_e^n on S_{AF} is confirmed by our analysis of the training effect in NiO/Fe heterostructures.

The investigation of the training effect reveals some information about the AF pinning system. Large training effects are observed in exchange-bias systems with polycrystalline antiferromagnets. In such systems the irreversible relaxation processes can even be thermally activated by switching of the AF grains. Such a model which shows significant analogies with the description of superparamagnetism has been developed [97]. As pointed out above, in exchange-bias systems which are based on single-crystalline antiferromagnets training effects can also be observed. This is again an indication for a limitation of the AF correlation length which originates from metastable AF domains. On the other hand, the absence of a training effect is a strong hint at alternative mechanisms providing AF interface magnetization which is unaffected by the magnetization-reversal of the FM top layer. The piezomagnetism discussed in Sect. 4.3.1 is a promising candidate for such a mechanism. In this case, where the exchange-bias heterostructures are based on the pure and diluted rutile-type antiferromagnets $Fe_{1-x}Zn_xF_2$ where the built-in shear-stress gives rise to a constant magnetic moment which is completely unaffected by the reversal of the exchange-coupled FM top layer.

References

1. C. Leighton, J. Nogués, B.J. Jönsson-Åkerman and I.K. Schuller: Phys. Rev. Lett. **84**, 3466 (2000)
2. S.S.P. Parkin and V. Speriosu: *Magnetic Properties of Low-Dimensional Systems, Vol. 2*, (Proc. Phys. **50**) (Springer, Berlin 1990),110
3. W.H. Meiklejohn and C.P. Bean: Phys. Rev. Lett. **102**, 1413 (1957)
4. W.H. Meiklejohn and C.P. Bean: Phys. Rev. **105**, 904 (1956)
5. W.H. Meiklejohn: J. Appl. Phys. **33**, 1328 (1962)
6. E.C. Stoner and E.P. Wohlfarth: Trans. R. Soc. (Lond.) A **240**, 599 (1948)
7. Joo-Von Kim, R.L. Stamps, B.V. Mc Grath and R.E. Camley: Phys. Rev. B **61**, 8888 (2000)
8. C. Tsang, N. Heiman and K. Lee: J. Appl. Phys. **52**, 2471 (1981)
9. A.P. Malozemoff: Phys. Rev. B **35**, 3679 (1987)
10. N.J. Gökemeijer, T. Ambrose and C.L. Chien: Phys. Rev. Lett. **79**, 4270 (1997)
11. E. Arenholz, K. Starke, G. Kaindl and P.J. Jensen: Phys. Rev. Lett. **80**, 2221 (1998)
12. J.S. Jiang, G.P. Felcher, A. Inomata, R. Goyette, C.S. Nelson and S.D. Bader: J. Vac. Sci. Technol. A **18**, 1264 (1999)
13. D. Mauri, E. Kay, D. Scholl and J.K. Howard: J. Appl. Phys. **62**, 2929 (1987)
14. X.W. Wu and C.L. Chien: Phys. Rev. Lett. **81**, 2795 (1998)
15. R.L. Stamps: J. Phys. D: Appl. Phys. **33**, R247 (2000)
16. P. Miltényi, M. Gierlings, J. Keller, B. Beschoten, G. Güntherodt, U. Nowak and K.D. Usadel: Phys. Rev. Lett. **84**, 4224 (2000)
17. R. Jungblut, R. Coehoorn, M.T. Johnson, J. van de Stegge and R. Reinders: J. Appl. Phys. **75**, 6659 (1994)
18. U. Nowak, A. Misra and K.D. Usadel: J. Appl. Phys. **89**, 7269 (2001)
19. Ch. Binek, A. Hochstrat and W. Kleemann: J. Magn. Magn. Mater. **234**, 353 (2001)
20. S. Riedling, M. Bauer, C. Mathieu, B. Hillebrands, R. Jungblut, J. Kohlhepp and R. Reinders: J. Appl. Phys. **85**, 6648 (1999)
21. J. Nogués and I.K. Schuller: J. Magn. Magn. Mater. **192**, 203 (1999)
22. A. Hochstrat, Ch. Binek and W. Kleemann: Phys. Rev. B **66**, 092409 (2002)
23. W. Kleemann: Int. J. Mod. Phys. B **7**, 2469 (1993)
24. A.P. Malozemoff: Phys. Rev. B **37**, 7673 (1988)

25. B. Kagerer, Ch. Binek and W. Kleemann: J. Magn. Magn. Mater. **217**, 139 (2000)
26. H. Xi and R.M. White: Phys. Rev. B **61**, 80 (2000)
27. Ch. Kittel: *Einführung in die Festkörperphysik* (Oldenbourg, München 1983)
28. C. Mathieu, M. Bauer, B. Hillebrands, J. Fassbender, G. Güntherodt, R. Jungblut, J. Kohlhepp and A. Reinders: J. Appl. Phys. **83**, 2863 (1998)
29. P.J. van der Zaag, A.R. Ball, L.F. Feiner, R.M. Wolf and P.A.A. van der Heijden: J. Appl. Phys. **79**, 5103 (1996)
30. D.P. Belanger and A.P. Young: J. Magn. Magn. Mater. **100**, 272 (1991)
31. D. Bertrand, F. Bensamka, A.R. Fert, J. Gelard, J.P. Redoulès and S. Legrand: J. Phys. C **17**, 1725 (1984)
32. H. Lueken: *Magnetochemie* (Teubner, Stuttgart 1999)
33. Landolt Börnstein: *Non-Metallic Inorganic Compounds Based on Transition Elements, New Ser. III, 27, Subv. j1* (Springer, Berlin 1994)
34. B. Kagerer, Ch. Binek and W. Kleemann: J. Magn. Magn. Mater. **217**, 139 (2000)
35. S. Maat, K. Takano, S.S.P. Parkin and E.E. Fullerton: Phys. Rev. Lett. **87**, 087202 (2001)
36. N.C. Koon: J. Appl. Phys. **81**, 4982 (1997)
37. N.C. Koon: Phys. Rev. Lett. **78**, 4865 (1997)
38. R.P. Michel, A. Chaiken and C.T. Wang: J. Appl. Phys. **81**, 5374 (1997)
39. H.P. Rooksby: Acta Crystallogr. **1**, 226 (1948)
40. J. Nogués, T.J. Moran, D. Lederman, I.K. Schuller and K.V. Rao: Phys. Rev. B **59**, 6984 (1999)
41. T.J. Moran, J. Nogués, D. Lederman and I.K. Schuller: Appl. Phys. Lett. **72**, 617 (1998)
42. P. Miltényi, M. Gruyters, G. Güntherodt, J. Nogués and I.K. Schuller: Phys. Rev. B **59**, 3333 (1999)
43. J. Nogués, D. Lederman, T.J. Moran and I.K. Schuller: Appl. Phys. Lett. **68**, 3186 (1996)
44. B. Kagerer: Senkrechte Anisotropie und exchange bias magnetischer Heteroschichten. Diploma Thesis, Gerhard-Mercator-Universität, Duisburg (1999)
45. U. van Hörsten: *Analysis of Small Angle X-ray Diffraction Data by Computer Simulation* (Gerhard-Mercator-Universität Duisburg, unpublished)
46. J. Lohau, A. Carl, S. Kirsch and E.F. Wassermann: Appl. Phys. Lett. **78**, 2020 (2001)
47. R. Allenspach and A. Bischof: Phys. Rev. Lett. **69**, 3385 (1992)
48. J. Thomassen, F. May, B. Feldmann, M. Wuttig and H. Ibach: Phys. Rev. Lett. **69**, 3831 (1992)
49. Y. Shapira: Phys. Rev. B **2**, 2725 (1970)
50. A.S. Borovik-Romanov: Sov. Phys. JETP **11**, 786 (1960)
51. J. Kushauer, Ch. Binek and W. Kleemann: J. Appl. Phys. **75**, 5856 (1994)
52. J. Nogués, D. Lederman, T.J. Moran and I.K. Schuller: Phys. Rev. Lett. **76**, 4626 (1996)
53. W.B. Zeper, F.J.M. Greidanus, P.F. Carcia and C.R. Fincher: J. Appl. Phys. **65**, 4971 (1989)
54. C.-J. Lin, G.L. Gorman, C.H. Lee, R.F.C. Farrow, E.E. Marinero, H.V. Do, H. Notarys and C.J. Chien: J. Magn. Magn. Mater. **93**, 194 (1991)
55. P. Fumagalli, A. Schirmeisen and R.J. Gambino: Phys. Rev. B **57**, 14294 (1998)

56. D. Lederman, J. Nogués and I.K. Schuller: Phys. Rev. B **56**, 2332 (1997)
57. M.E. Lines: Phys. Rev. **156**, 543 (1967)
58. Ch. Binek, B. Kagerer, S. Kainz and W. Kleemann: J. Magn. Magn. Mater. **226–230**, 1814 (2001)
59. A.M. Goodman, H. Laidler, K. O'Grady, N.W. Owen and A.K. Petford-Long: J. Appl. Phys. **87**, 6409 (2000)
60. X. Portier, A.K. Petford-Long and A. de Morias: J. Appl. Phys. **87**, 6412 (2000)
61. Ch. Binek, Xi Chen, A. Hochstrat and W. Kleemann: J. Magn. Magn. Mater. **240**, 257 (2002)
62. I.S. Jacobs and P.E. Lawrence: Phys. Rev. **164**, 866 (1967)
63. J.F. Dillon, E. Yi Chen and H.J. Guggenheim: Solid State Commun. **16**, 371 (1975)
64. N. Papanicolaou: Phys. Rev. B **51**, 15 062 (1995)
65. J.T. Randall and M.H.F. Wilkins: Proc. R. Soc. A **184**, 347, 364, 391 (1945)
66. G.F. Garlick and M.H.F. Wilkins: Proc. R. Soc. A **184**, 408 (1945)
67. K. Binder and J.D. Reger: Advan. Phys. **41**, 547 (1992)
68. F.U. Hillebrecht, H. Ohldag, N.B. Weber, C. Bethke and U. Mick: Phys. Rev. Lett. **86**, 3419 (2001)
69. I.E. Dzialoshinskii: J. Exptl. Theoret. Phys. (U.S.S.R.) **33**, 807 (1957)
70. A. Hochstrat: Herstellung und Untersuchung magnetischer Heteroschichtsysteme. Diploma Thesis, Gerhard-Mercator-Universität, Duisburg (2001)
71. P. Miltényi, M. Gierlings, M. Bamming, U.May, G. Güntherodt, J. Nogués, M. Gruyters, C. Leighton and I.K. Schuller: Appl. Phys. Lett. **75**, 2304 (1999)
72. J. Mattsson, C. Djurberg and P. Nordblad: J. Magn. Magn. Mater. **136**, L23 (1994)
73. C. Hou, H. Fujiwara, K. Zhang, A. Tanaka and Y. Shimizu: Phys. Rev. B **63**, 024411 (2000)
74. R.B. Griffiths: Phys. Rev. Lett. **23**, 17 (1969)
75. Ch. Binek and W. Kleemann: Phys. Rev. B **51**, 12888 (1995)
76. Ch. Binek, S. Kuttler and W. Kleemann: Phys. Rev. Lett. **75**, 2412 (1995)
77. K. Binder: *Phase Transitions and Critical Phenomena, Vol.3*, ed. by C. Domb and J.L. Lebowitz (Academic, London 1983) p. 1
78. M. Pleimling and W. Selke: Eur. Phys. J. B **1**, 385 (1998)
79. M. Pleimling: Comput. Phys. Commun. **147**, 101 (2002)
80. C. Ruge, S. Dunkelmann and F. Wagner: Phys. Rev. Lett. **69**, 2465 (1992)
81. H.W. Diehl and M. Shpot: Phys. Rev. Lett. **73**, 3431 (1994)
82. L. Onsager: Phys. Rev. **65**, 117 (1944)
83. U. Nowak, A. Misra and K.D. Usadel: J. Magn. Magn. Mater. **192**, 203 (1999)
84. D. Paccard, C. Schlenker, O. Massenet, R. Montmory and A. Yelon: Phys. Status Solidi **16**, 301 (1966)
85. C. Schlenker, S.S.P. Parkin, J.C. Scott and K. Howard: J. Magn. Magn. Mater. **54**, 801 (1986)
86. K. Zhang, T. Zhao and M. Fujiwara: J. Appl. Phys. **89**, 6910 (2001)
87. S.G. te Velthuis, A. Berger and G.P. Felcher: J. Appl. Phys. **87**, 5046 (2001)
88. W. L. Roth: J. Appl. Phys. **31**, 2000 (1960)
89. W. Kleemann, F.J. Schäfer and D.S. Tannhauser: J. Magn. Magn. Mater **15**, 415 (1980)

90. H. Ohldag, A. Scholl, F. Nolting, S. Anders, F.U. Hillebrecht and J. Stöhr: Phys. Rev. Lett. **86**, 2878 (2001)
91. M. Ohldag, T.J. Regan, J. Stöhr, A. Scholl, F. Nolting, J. Lüning, C. Stamm, S. Anders and R. L. White: Phys. Rev. Lett. **87** 247201 (2001)
92. H. Matsuyama, C. Haginoya and K. Koike: Phys. Rev. Lett. **85**, 646 (2000)
93. J.A. Borchers, Y. Ijiri, D.M. Lind, P.G. Ivanov, R.W. Erwin, A. Quasba, S.H. Lee, K.V. ODonovan and D.C. Dender: Appl. Phys. Lett. **77**, 4187 (2000)
94. A. Hubert and R. Schäfer: *Magnetic Domains: The Analysis of Magnetic Microstructures* (Springer, Berlin, Heidelberg, New York 1998) pp. 108–155
95. S. Blundell: *Magnetism in Condensed Matter* (Oxford University Press, New York 2001)
96. F. Radu, M. Etzkorn, R. Siebrecht, T. Schmitte, K. Westerholt and H. Zabel: J. Magn. Magn. Mater. **240**, 251 (2002)
97. M.D. Stiles and R.D. McMichael: Phys. Rev. B **60**, 12 950 (1999)

5 Summary

By selecting various timely topics it has been the aim of this review to point out the significance of Ising-type antiferromagnets to gain deeper insight into basic problems of statistical physics on the one hand and to understand the mechanism of exchange-bias in magnetic heterostructures on the other hand. As a guiding principle, simplicity of the crystalline and magnetic structure of the investigated AF compounds has been chosen in order to realize prototypical behaviors from various points of view.

For example, the layered structure of $FeCl_2$ is at the origin of a hierarchy of direct and indirect exchange interactions. Together with the strong crystal field-induced single ion anisotropy it gives rise to FM 2d Ising behavior well above T_N. This property has been used in order to get experimental access to the statistical theory of Lee and Yang from the analysis of high-field magnetization data.

Moreover, the simplicity of the spin structures and the respective strong uniaxial anisotropy make $FeCl_2$ and, in particular, $Fe_{1-x}Zn_xF_2$, prototypical AF pinning systems of exchange-bias heterostructures. The susceptibility of the AF pinning layer controls various types of coupling phenomena which emerge from the mutual interaction between the AF and FM constituents. In order to fabricate perpendicular exchange-bias systems FM Co/Pt multilayers with perpendicular anisotropy have been deposited on top of the uncompensated surfaces of the Ising antiferromagnets. Simplicity of the involved spin structures is achieved due to the minimized number of spin degrees of freedom.

This allows for a description of the experimental findings where a generalized Meiklejohn–Bean approach is the starting point for the interpretation of, e.g., the freezing field and temperature dependence of the exchange-bias field. The necessity of AF interface magnetization is an important consequence of the simple description of exchange-bias.

In view of this fact, it is a challenging task to explore the origin of exchange-bias in heterostructures with compensated AF interfaces. According to the recently proposed domain-state model the AF interface magnetization originates from random-field-induced domains, which are created on cooling a diluted antiferromagnet in a field. In addition, however, piezomagnetism has to be taken into account as a mechanism which provides magnetization at compensated AF interfaces. Again, the rutile structure of FeF_2 gives rise to prototypical piezomagnetic behavior. The onset of a piezomagnetic moment below the Néel temperature is clearly evidenced in the $Fe_{1-x}Zn_xF_2/Fe$ heterostructure. It is supposed to support exchange-bias in heterostructures based on diluted systems and may be even of major importance in the case of pure antiferromagnets where the presence of random-field domains is not obvious.

The above list of problems involving insulating ionic AF compounds demonstrates their crucial role for the experimental realization of classical spin systems with localized magnetic moments. Hence, it is certainly not exaggerated to expect that Ising-type antiferromagnets will remain important model systems linking between applied experimental research and theoretical investigations on magnetism and –more generally– statistical physics.

Index

Springer Tracts in Modern Physics

Springer Tracts in Modern Physics

Printed in the United States
144981LV00003B/145/P

9 783540 404286